Want to Ace Organic Chemistry?
Even if you currently have an F?

Get the Ultimate Vault of **Proven Study Tools** to Ace Your Organic Chem Exams
You get a 50% discount just for buying this book.

✓ *"I went from a D to an A. Thank you so much!".*

✓ *"Easily explained some difficult concepts. Visual aids REALLY helped. Easy to understand and follow presenter."*

✓ *"Your Organic I videos helped me so much for the first semester. You guys are doing a great thing!!"*

With your enrollment you get:

- Videos, flashcards, eBooks, mini-movies, practice exams, and MUCH more proven to get you results.
- Weekly emails from your personal Sherpa, telling you what to study with links to find it, save you study time.
- Study plan with links to the material, based on the grade you want.
- 24/7 access anytime, anywhere on any device, to study on your time.
- 24/7 support to ensure your success
- Material is continually created to give you even more to help.

There is no risk to try it:

- Reason 1: Cancel anytime 24 hours, no questions asked.
- Reason 2: You get to keep the eBooks and study guides if you do cancel. The books go for $30 on Amazon.
- Reason 3: We take care of our students. We succeed when you succeed.

No Risk Discount Enrollment at
www.AceOrganicChem.com/special.html

Table of Contents

#28- It takes alkynes to make the world go 'round (and to react with epoxides)

#29- Four organometallics to rule them all

#30- Diels-Alder Reaction-Part 1: An introduction

#31- Let's go retro: Retrosynthetic analysis

#32- Overall EAS strategy

#33- EAS strategy: Turn a meta into ortho/para and vice-versa.

#34- EAS strategy: Conversion of alkyl groups to carboxylic acids.

#35- EAS strategy: In football, you need good blockers. SO_3 and X are our blocking groups

#36- EAS strategy: Long chain alkyl groups from Wolff-Kishner or reduction

#37- EAS strategy: Substituted toluenes came from toluene. Duh

Chapter 4: Helpful Reactions, Hints, Clues

#38: Want to impress your instructor? Start dropping names.

#39- Enantiomers are siblings, diastereomers are cousins which makes rotomers your dingy aunt.

#40- Benzylic and allylic leaving groups are better than they look.

#41- Before you die, you see the [aromatic] ring.

#42- "Eject! Eject! Eject!" Carbonyls with an ejectable group are added to twice.

#43- Hint: "hv" means a radical is brewing.

#44- T-butyl groups on cyclohexane means no ring flips….kinda.

#45- Like Jerry Lewis asking for money, Lewis acids are always asking for electron density.

#46- H_2SO_4 and HNO_3: The good-cop/bad-cop of nitrations.

#47- The Incredible Bulk: Bulky bases.

#48- Hoffman vs. Zaitsev: The elimination.

#49- Dude, where's my carbocation?

#50- Free Radical Halogenation: The Molecular Handle.

#51- My Three Reductions: Reduction Methods For Alkynes

#52- Is a Halogen Squatting on Your Molecule? Removing the unwanted halogen.

#53- You don't want a "D" on your transcript, but you might want one on your molecule.

#54- Diels-Alder Part 2: Regiochemistry and stereochemistry

#55- Diels-Alder Part 3: What makes a good Diels-Alder reaction

#56- It's an urban legend that S_N1 reactions give *completely* racemic products.

#57- "What do you 'Amine' you're getting rid of me?"--How to eliminate an amine.

#58- "Why didn't you say were sendin' the wolf?"—The Wolff-Kishner reduction

#59- It takes a Baeyer-Villiger to raise a child

#60- Symmetric diols came From the Pinacol reaction

#61- Convert a nitro group into the aldehyde or ketone.

#62- The McMurry reaction: A "reverse" ozonolysis.

#63- Role reversal: Nitrile hydration and the Letts Nitrile Synthesis

#64- Role reversal: Dihydroxylation and the Corey-Winter Olefination

#65- Role reversal: Halogens and carboxylic Acids

#66- Musical double bonds: Using catalytic I_2.

#67- Just like the Secret Service, protecting groups are chemical bodyguards

#68- Don't hydrogenate that aryl ring! You'll kill us all!

#69- It's as easy as 1,2 addition. Or is It 1,4 addition?

#70- Here's to your annulation, Mrs. Robinson: Musings on the Robinson Annulation.

#71- Kinetic vs. thermodynamic enolates: You choose the winner of this fight.

#72- "Where's Aldol?": Making an aldol reaction work for you.

#73- Acetoacetic Ester Synthesis: What you need to know in one page or less.

#74- β-keto esters came from a Claisen Condensation (NOT the Claisen Rearrangement)

#75- Convert ketones to epoxides fast

#76- Two ways to open one epoxide

Chapter 5: Spectroscopy

#77- SODAR is not just a drink mixer anymore

#78- Splitting should be automatic, at least with NMR

#79- O, M, P from the aromatic region of an NMR

#80- Everything You Ever Wanted to Know About a ^{13}C NMR, But Were Afraid to Ask

#81- There are Only Four Important IR Peaks.

#82- Check Out the Cleavage on That Molecule

#83- The Nitrogen Hint (Not a Rule)

Chapter 6: Study Tips and Suggestions

#84- Are You a Learner Like Socrates or a Memorizer Like a Super Computer?

#85- What a Tangled Web We Weave.

#86- Be a Chatty Patty and Talk Out Your Reactions.

Introduction: How to use this book effectively.

There are many things that this book can be for you. But first, we should discuss what this book is not. This is not meant to be a comprehensive treatise on organic chemistry. Quite the opposite actually, this book is meant to be a short supplement to the organic chemistry text that you are already using in your class. **There are some very important concepts in organic chemistry that this book will not even touch**. That does not mean that they will not be on your exam; it just means that we were trying to hit on topics where there are tricks that can be learned to make that topic easier.

That being said, here is how to use this book most effectively:

1) **Try to understand the topic from your text first**, then refer to this book as a supplement to see if there is an easier way to learn it.

2) **Determine what grade you want then study to get that grade.** All too often, the students who only want a "C" spend way too much time studying the hardest concepts of organic chemistry and don't master the material that will get them that C. This book breaks each tip down by difficulty:

Ex:
> **Orgo II Difficulty: ★**

> Three Stars: Material for the "A" students. These are the most difficult concepts or tricks. Some of them may not even be taught in your class. These are meant for the true organic masters.

> Two Stars: Most undergraduate students will be taught these concepts. The tricks and tips that we will give you with these concepts are meant for almost all students. Students who want a grade of "C" or better should become familiar with these concepts.

> One Star: The very basics and should be learned inside out in order to ensure a good grade in organic chemistry. If you are having trouble with these concepts, it is going to be a very long semester for you.

3) **If one of the tricks does not make sense to you, _DON'T_ use it.** Your mission is for you to learn organic chemistry, whatever way you can. If one of these tricks does not make sense to you, just learn it another way.

That being said, this book will give you some nice material that you may not easily find in other places. Use it to your advantage, study hard and you will get the grade you want.

To my lovely (and understanding) wife, Katie Bug, Moochie Monster, T-Rex and Abidale.

<u>Chapter 1</u>: Setting Yourself up for Success

#1- How to get an "A" (or just pass) Organic Chemistry

"I feel like I have a brain roadblock where things just don't click."
"My mind is turning to mush"
"Organic chemistry is too depressing"

These are just a few quotes that I pulled off of Reddit on day. Let's face it, organic chemistry is tough. But more troubling is the number of students that become discouraged. However tough organic chemistry may be, it is not impossible. And more than that, it can be rewarding insomuch as it could be a giant course in your career path or could make or break your medical school application.

So, if organic chemistry is important, then how can we make it easier to learn and study for? The easiest answer is to ask the men and women who teach it daily. Surprisingly enough, I searched and searched and could not find any comprehensive guide or survey of what professors thought the best way to study is. So, we did it ourselves. We asked a bunch of organic chemistry professors a bunch of questions about their students, the course and the most effective ways to study. We compiled all the results and were a bit surprised at some of the responses.

First, the bad news: The professors we surveyed advised us that on average greater than 40% of their students want to go to medical school. This can be challenging for several reasons. Aside from the fact that pre-meds have the reputation (deservedly or not) of being neurotic cut-throat curve-wreckers, some professors make organic chemistry a "weed em' out" class to get rid of anyone from the pre-med program that isn't going to make it. Unfortunately, this could mean that you get caught in the weeding out.

More bad news: 77% of the professors told us that students need to study for their class more than eight hours per week, and almost 50% said it needs to be more than 10 hours per week. That is a lot of time to spend on one course. Moreover, 10 hours per week is if it clicks with you. If the material does not come naturally to you, it will take even more than just 10 hours per week.

These two factors come together and result in almost 25% of students not completing first semester and moving to organic II.

Now for some good news: Over 70% of the professors in our survey said that studying outside of class is the most important thing a student can do to succeed. Surprisingly, this is far greater than the 23% that said making it to lecture every time was more important than outside study. So, if studying outside of class is important, what or how should a student study?

PRACTICE EXAMS ARE THE KEY! The #1 answer to that question is practice exams. Almost 90% of the professors we surveyed said this was the best and most efficient way to study for o-chem. This was an overwhelming majority. Professors lumped practice exams into larger category of practice problems, which they told us are something students should spend a lot of their study time on. We like a couple of resources for finding applicable practice exams. First, is your professor's own course website, where they tend to post old exams for studying. This is nice because not only are they easy to find, but students can also get a good idea of the topics their professor likes and how she might ask questions. Moreover, professors can sometimes be lazy and reuse questions, which means you will have a jump on that problems because you will have already seen in on a practice exam.

LEARN LOTSA MECHANISMS. The second most popular way among our professors for studying is by learning as many reaction mechanisms as you can. This will not only help you to learn the reaction itself by seeing it again, but it will reinforce why the reaction works, which will assist you if you must figure out a reaction or mechanism you might have never seen before. There are a couple of ways we suggest learning mechanism, including Amazon books on the topic. Our favorites are *Pushing Arrows*, Ace Organic Chemistry Mechanisms with EASE, and *The Art of Writing Reasonable Organic Chemistry Mechanisms* (Advanced), which are all available on Amazon. Internet resources can be found by Googling "Organic Chemistry Mechanisms".

VIDEO, VIDEO, VIDEO. The third most popular way to study, according to our professors, is by watching video. We observe that there are three types of organic chemistry videos on the web. The first are full lectures, in which a full semester of videos can be 30-50 hours long. These are usually webcasts of a professor's entire semesters lectures, put into podcast form and available to anyone. iTunes or a university website are the best places to find these. The second type of summary videos, which can be anywhere from 6-14 hours in total length to cover a semester. These are nice because they condense the material and are a bit easier to digest, but still cover all the major topics. YouTube, iTunes, or organic chemistry help websites are a great place to find these. Finally, specific topic videos are the third type of video you will find. These are 10-25 minutes long and focus on specific topics or reactions in organic chemistry. These are nice for students who are doing pretty well in the class but might be struggling with a specific area or reaction. YouTube is a great place to start looking for these.

Professors were very certain about what not to study: Almost every professor we surveyed said that students should not just sit down and read the textbook again. They felt this was just not productive and that there were better ways to study than reading a textbook that you should have already seen once during your lecture.

We also asked professors what the best resource on the web was for their students. Surprisingly, the two most popular answers were "my website" and "nothing", which said to us that professors are partial to their own website and not much else more.

Finally, we asked professors which accessories they liked best. By far, a majority said modeling kits were a great study aide because they allow students to better visualize molecular structure and reactions. We have created a high-quality model kit with a 2-hour DVD to show you all the tricks to learning with it. Just click here to learn more about it.

There you have it. An exhaustive survey on the best ways to really crush this course. Remember, it is NOT impossible. You can do this.

#2- Study in packs—there is safety in numbers.

This reminds me of my favorite video on YouTube (QR code below). You are a pack of wildebeest, just chilling out by the water, trying to score a good grade in organic chemistry. But you are being hunted by pride of hungry lions (your professors) who would like nothing better to make a quick snack of the weakest one of you. After crouching in the brush, the lions suddenly pounce (pop quiz) and grab a hold of the smallest one of you (the student with the hardest course load).

Two things can happen at this point: Either the rest of the pack of wildebeest will cut their losses and try to save themselves or they can go back and heroically battle the lions to rescue their fallen colleague. I am not going to ruin the video if you have not already viewed it, but I think you already know what happens.

More than just helping others, studying in packs provides several other benefits:

1) Studies have shown over and over that studying in groups directly leads to higher grades for all participants.

2) Studying in groups is generally more enjoyable for people, which leads to more time spent on the subject.

3) If you are weaker in one area of the course, you have the opportunity to have a peer explain it to you. Many students are more likely to understand a peer's explanation over a stuffy professor's.

4) If you are stronger in one area of the course, you will strengthen your overall understanding of chemistry by teaching it to someone else.

Of course, when you are choosing study partners on the Serengeti, you need to be very careful to stay away from the jackals. These are the students that are more parasite than human and will just leach off of your talents. They will come to study sessions unprepared and expect you to teach the entire course to them. They are more Succubus than man and will not help you much. We suggest finding study partners that are interested in a good grade and are willing to put in the time necessary to achieve high marks in the course.

The Battle at Kruger:

#3-Take the course at a more opportune time.

A useful strategy that many students have used is to take chemistry in the "off season". This means that they have taken Organic I in the spring and then followed with Organic II in the next fall. This presents several advantages, including that since you are "out of sequence" the class sizes will most likely be smaller; this will usually provide for a better learning environment.

Many will ask if they should take it in the summer. This is not an easy "yes or no" question, and definitely depends on the student. Here are some of the considerations for summer classes:

The Good:

1. It is only usually 5 weeks long.

2. If you are not working that summer, it is much better than just sitting around doing nothing.

3. If you are not majoring in chemistry and don't want to go to medical school, it is a great way to get organic chemistry out of the way quickly.

The Bad:

1. If you ARE majoring in chemistry, it is very easy to forget everything that you learned in the class because you crammed it all into 5 weeks.

2. Classes are usually at least 3 hours per day, plus homework every night and an exam once a week. This can be overwhelming.

The Ugly:

1. If you get a bad professor, keep the hemlock close.

#4- Flashcards can be your new BFF.

We are not huge fans of simply memorizing information. This leads to students memorizing, regurgitating, forgetting, and not actually learning a subject. That being said, flashcards get a bad rap sometimes because some equate them with memorization. While this is the case sometime, it doesn't have to be. Flashcards, in their most basic form, are just a way to ask students a question and have easy access to the answer. Here are some things to think about with respect to flashcards.

1) Making your own flashcards is the best way to learn the material because they are specific to your course and because you learn as you create them.

2) Flashcards don't have to be one-word answers. You can put a reaction on the back, or a mechanism, or really anything you want to learn.

3) Flashcards can be physical or on a website. Google "organic chemistry flashcards" to find some really good (free) sites.

4) You can purchase some on Amazon too. This isn't bad because they give you a different perspective, are pretty durable, and are always going to be accurate information.

Learn to love your flashcards and you will go a long way in this course.

#5- Don't over-emphasize the lab section.

One of the biggest complaints that most TAs hear is that the laboratory section of organic chemistry is a lot of work for just one credit. There are no two ways about it…this is a true statement. Think about it. If you were to do everything most TAs require for the lab, you would spend at least one hour prior to the lab preparing for it, three hours in the lab itself, and two hours analyzing your data and writing your report. This equals six hours of work per week for one lousy credit. Compare this to your organic chemistry lecture class, which is usually three hours of class for three credits.

We can't change the system, but we can learn how to work it to our advantage. With respect to the pre-lab preparation, one trick we have found that works well is to get together with a group of lab mates and split the work on the lab preparation by alternating who does the lab prep for that week. Some TAs want to you to have the MSDS sheets for all chemicals to be used, in addition to answering pre-lab questions about what you are going to do in lab that day. By alternating who does that each week, you could save hours of time that could be devoted to the lecture portion instead.

The report to be written after each lab exercise is another place that many students waste countless hours. Remember that the most important part of the lab is to do the calculations correctly. It does not matter that your data is crappy, it matters what you do with that crappy data. Second, do not spend a ton of time on the narrative portion. Concisely write what you did and leave it. Don't fool yourself into thinking that your TA is going to go over this with a fine-tooth comb.

Many will find this to be a controversial statement, but we have found that a "B" in the lab section is not going to preclude you from going further in science, whether that be to a graduate degree or medical school. The one credit from the lab section does not factor that heavily into your "science GPA" or overall GPA. If you could spend three hours less per week on the lab and focus that time into studying the lecture portion, you are much more likely to see a tangible benefit down the road. Further, would it be worth spending that extra three hours on the lab portion if it were only going to raise your lab grade a little bit? (For example, from a "B" to a "B+") This is even more true if all you want is to pass the lab section and don't care about the grade.

Do not mistake what we are saying here, as the lab portion can be a fun and rewarding experience. We are only advocating that for some students it may be helpful to not over-emphasize a class that is only worth one credit.

<u>Chapter 2</u>: Introductory Concepts

#6- Memorize your nomenclature and essential vocabulary.

Learning organic chemistry is like trying to work in a foreign country; if you don't know the language, it is going to be very difficult to learn how to do your job. Imagine that you have just been transported to the mythical country of "ochemia", a small island nation in the south Pacific, where your job is to write chemistry reactions.

Frequently, in a chemistry lecture, professors start tossing out strange organic chemistry terms far too quickly. Because students aren't fluent in "ochemia" yet, they need to translate each word in their head to understand what the instructor has just said. By the time this mental translation is done, the student has just missed the next sentence and has lost half of the lecture. Our goal is to get as fluent as we can in the language of chemistry as quickly as we can. Here are <u>some</u> terms it will be helpful to memorize so that you don't have to do a mental translation when you hear them:

Meth = 1

Eth= 2

Prop = 3

But = 4

Pent = 5

Hex = 6

Hept = 7

Oct = 8

Non = 9

Dec = 10

Nucleophile = has electrons, has a negative or partial negative charge

Halogen = F, Cl, Br, I

Aprotic solvents = do not contain OH or NH bonds

Protic solvents = contain OH or NH bonds

Lewis Acid = electron acceptor

Lewis Base = electron donor

Carbonyl group = (C=O)

Cis = same side of a double bond or ring

Trans = opposite sides of a double bond or ring

Electrophile = wants electrons, has a positive or partial positive charge

#7-Wouldn't formaldehyde by a different name smell just as sweet? The difference between common names and IUPAC nomenclature.

We all have nicknames. This is not Earth-shattering information, and chemical compounds are no different. Your job is to determine which compound is being discussed when your professor refers to it by three different names.

As shown below, the same compound can have a couple of different names. There are some compounds that we never use the systematic name for. 1,3,5-cyclohexatriene would be one of those, as we never use that name; we always refer to it as benzene. Further, most students are not familiar with 2-propanone, but most of us know what acetone is.

So, what is the right strategy to use so as not to confuse these compounds? We believe that you can _almost_ never go wrong using the IUPAC name of a compound. The reason the IUPAC nomenclature was developed was to eliminate confusion with informal names of chemical compounds. That being said, you should be aware of some common names also. With substituent groups and many large molecules, it will be much easier to refer to the compound by its common name. Below, we give some molecules that you should know by their common names:

Structure	IUPAC name	Common name
	2-propanol	Isopropanol
HC≡CH	Ethyne	Acetylene
	2-propanone	Acetone
	Ethanoic Acid	Acetic Acid
	2-methyl 2-propanol	t-butyl alcohol

Take Home Message: Use the IUPAC name for almost all molecules, except for a few that you should know the common names.

#8- Know the "normal" state for common organic atoms.

Structures of molecules can be difficult to piece together at first when you are just starting in an organic chemistry class. One of the tricks that can greatly help with this is to know the uncharged or "normal" state for atoms that are commonly found in organic molecules. Here is a table of the most common of those:

In an UNCHARGED state:

  - C has four bonds (and no lone pairs)
  - N has three bonds (and one lone pair)
  - Halogens have one bond (and three lone pairs)
  - O has two bonds (and two lone pairs)
  - H has one bond (and no lone pairs)

Three more rules:

-     C, N, O are central atoms, meaning that they will almost always be in the middle of your molecule, never as a terminal atom (IF they are neutral)

-     H and halogens are almost always terminal atoms, meaning that they will only have one bond and be at the ends of molecules.

-     With the exception of H, atoms in group I & group II are only counterions (+1 or +2 charge and not involved in resonance).

Remember, these rules are for when the atom is uncharged; this *does not* apply to charged atoms. For example, a carbocation (a positively charged carbon atom) will have only three bonds with no lone pairs while a carbanion (a negatively charged carbon atom) will have three bonds with one lone pair.

Notice that all of these atoms still follow the octet rule. However, beware of atoms that do not follow the octet rule, as phosphorus is an example of an atom that can have more than an octet of electrons. Shown below is triphenylphosphine oxide, a byproduct of the Wittig reaction.

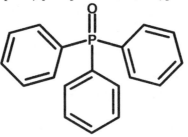

Elements with open d-subshells, like phosphorous and sulfur, do not always follow the octet rule. More examples of this are SF_6 and PCl_5. However, carbon, nitrogen and oxygen will follow the octet rule.

Bonus: "Wait, there was a lone pair there?"

Lone pairs are like the anti-lock brakes on your car. Sometimes you don't think about them until you need to. Lone pairs are easy to forget, and many times, we won't even include them when we draw a molecule. But to a new student in organic chemistry, this can cause plenty confusion. Look at the molecules below:

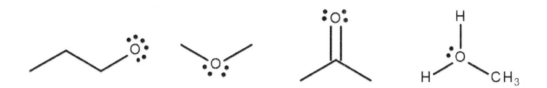

Without showing either the atomic charge on the oxygen, or the number of lone pairs, we might have a bit of trouble telling what is really going on with these. Moreover, oxygen, nitrogen, and carbon can all be positive, negative or neutral depending on the number of bonds and lone pairs.

And speaking of that, it is the number of bonds and lone pairs which determine the formal charge on an atom. The formula, which you might remember from general chemistry, is as follows:

FORMAL CHARGE = Valence electrons − [lone pairs + bonds]

Bonds get treated like one electron around the atom because the two electrons in the bond are shared between two atoms. Once we apply the formula above, it becomes much simpler to determine the formal charge.

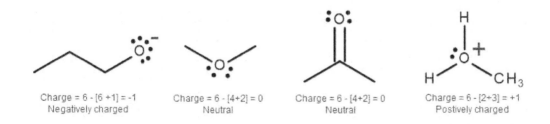

| Charge = 6 - [6 +1] = -1 | Charge = 6 - [4+2] = 0 | Charge = 6 - [4+2] = 0 | Charge = 6 - [2+3] = +1 |
| Negatively charged | Neutral | Neutral | Postively charged |

Those of you that retain information best by understanding it, you should remember formal charge is # of valence electrons minus the number of electrons actually circling the atom. For those of you that are all about the memorization, here is a chart for you to commit to memory.

Atom	Lone Pairs	Bonds	Charge	Example
Carbon	0	4	0	Alkanes
Carbon	1	3	-1	Carbanions
Carbon	0	3	+1	Carbocations
Nitrogen	0	4	+1	Nitronium ions
Nitrogen	1	3	0	Amines
Nitrogen	2	2	-1	Amides
Oxygen	1	3	+1	Hydronium ion
Oxygen	2	2	0	Ethers, alcohols
Oxygen	3	1	-1	Oxides

Remember here that lone PAIRS are two lone electrons.

Why do we care about this chart? Because the final charge of an atom can tell us a lot about the molecule's role in a reaction. For example, a negatively charged oxygen in a reaction is almost always going to act as a base or nucleophile whereas a positively charged carbon is most likely going to act add an electrophile in the reaction.

#9- Fischer projections are a black-tie affair.

Emil Fischer is considered by many to be the greatest organic chemist to ever live. His problem was that he created a way of looking at organic molecules that is very confusing to undergraduates. These structures are necessary to learn and are very helpful when looking at certain molecules (such as carbohydrates), but they are also very easy to jumble. This is because Fischer structures are drawn as crosses, which could lead one to erroneously think that the central carbon is flat, when it is actually still tetrahedral.

The easiest way to look at these is to think of them as bowties that have been strung together:

The "bowtie"

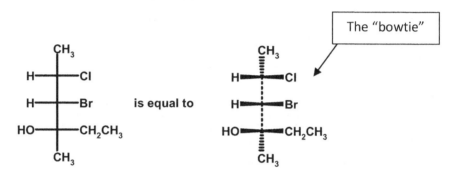

3-dimensionally speaking, the substituents that are on the sides of the structure are depicted at the end of the bowtie and are represented as "coming out of the paper". The backbone is composed of dashed lines, which are meant to represent that those portions "are going into the paper". This is now a much easier way to view these structures, as it is more apparent what area each substituent occupies.

The useful part of the bowtie projection is that it is now easier to assess the stereochemistry at each chiral center. It should be much easier to visualize that the bottom chiral center is "R". This was not as obvious when viewing the Fischer projection as a cross.

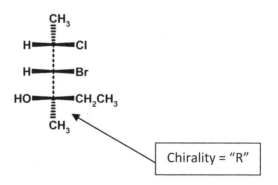

Chirality = "R"

#10- Is it really that "EZ"? Determining alkene geometry.

Way back when you were just an organic chemistry infant, you learned that alkenes existed and that they could be labeled either as a cis alkene or a trans alkene. It was nice, and you were happy not knowing that there were other, more sinister, alkenes out there that did not want to fall into such neat little categories. One of these diabolical olefins (another name for an alkene) is below:

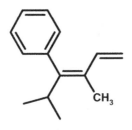

Further into your organic chemistry odyssey, you are going to worry about how you might synthesize this molecule. Right now, you are merely concerned with what to name it. Is this a cis or trans alkene? Unfortunately, cis and trans do not apply to more complex alkenes, so we need a new system of nomenclature. Enter the Cahn/Ingold/Prelog system for naming double bonds.

The system itself is simple: assign the atom connected to the olefin a priority based on its atomic number then determine whether the two highest priority groups are on the same side of the double bond, or on opposite sides. For example:

In the above example, we compare Cl to F and quickly determine the Cl is of greater priority based on its atomic number. When the same comparison is made for H and CH_3, we determine that C is the higher priority. Thus, the alkene above is an "E" alkene.

Further, there is a nice mnemonic used to remember which letter goes with which alkene:

- "Z" alkenes are on the "ZAME side"
- "E" alkenes are far "E-WAY" from each other.

While figuring out which type of alkene is present is easy when there are four different atoms connected to the alkene, we have to ask what would happen if there was a tie because there were two of the same atom on the alkene.

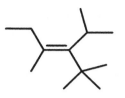

In the above alkene, we are posed with the problem of four carbon atoms bound to the double bond. Here, we can use the AceOrganicChem.com method for determining priority. For each substituent on the alkene, write out the atom attached directly to the alkene and each of the three atoms attached to it, in order of atomic number. For example, -CH$_3$ would be C (H, H, H), and -CH$_2$CH$_3$ would be C (C, H, H). If we were to say what we were writing, an ethyl group of C (C, H, H) would translate to "a carbon with one carbon and two hydrogens attached to it". Since they are already in the parentheses in order of atomic number, ties are easy to judge. If the first atoms in the parentheses are the same, just go to the second atom to break the tie. Thus, an isopropyl group with C (C, C, H) outranks an ethyl group with C (C, H, H)

Now, our molecule looks like this:

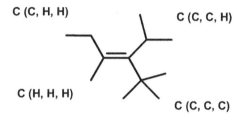

C (C, H, H)

C (C, C, H)

C (H, H, H)

C (C, C, C)

It can now easily be seen that this is an "E" alkene. This method can also be applied to other functionalities common found attached to olefins:

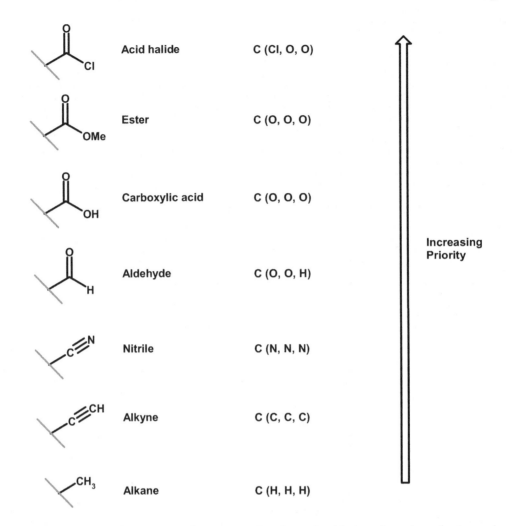

Acid halide	C (Cl, O, O)	
Ester	C (O, O, O)	
Carboxylic acid	C (O, O, O)	
		Increasing Priority
Aldehyde	C (O, O, H)	
Nitrile	C (N, N, N)	
Alkyne	C (C, C, C)	
Alkane	C (H, H, H)	

The trick to looking at these groups is to recognize that a double bond on the substituent is equal to two of that type of atom. For example, a cyano group would be equal to C (N, N, N). Another example of this would be as follows:

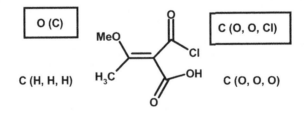

The only other trick to be aware of is a problem that has isotopes of the same atom. For example, deuterium is a higher priority than hydrogen because it is heavier. Therefore, the example below would be a "Z" alkene.

<u>Take Home Message</u>: When determining E/Z configuration of an alkene, use the C (X, X, X) method for determining priority, and remember that "Z" is the "ZAME side" and "E" is far "E-way" from the other group.

#11- Master Resonance.

Resonance is one of those issues that you will have to deal with for both semester I & II of organic chemistry. It is much better to have a solid understanding of it now, rather than have to worry about it later. The basic goal of resonance structures is to show that molecules can move electrons and charges onto different atoms on the molecule. This makes the molecule generally more stable because the charge is now delocalized and not "forced" onto an atom that does not want it.

Below are some handy rules of resonance. If you learn these and think about them when tackling different resonance problems, you will be able to handle whatever is thrown at you.

1) Know each atom's "natural state", see tip #8. You need to recognize what each atom generally looks like, in an uncharged state. This will help you to construct the Lewis Dot structure on which you will base your resonance structures. Remember that halogens and hydrogens are always terminal, meaning that are at the end of the molecule and only have one bond, and therefore, they will not participate in resonance.

2) Atom positions will not change. Once you have determined that an atom is bonded to another atom, that order will not change in a resonance structure. If they do change, it is no longer a resonance structure, but is now a constitutional isomer.

3) Check the structure you have created to make sure that it follows the octet rule. This will become much easier once you have a better handle on the "natural state" of atoms.

4) When two or more resonance structures can be drawn, the one with the fewest total charges is the most stable. In the example below, A is more stable than B.

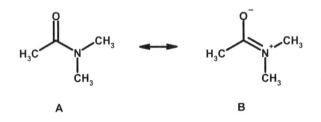

A   B

5) When two or more resonance structures can be drawn, the more stable has the negative charge on the more electronegative atom. In the example below, A is more stable than B.

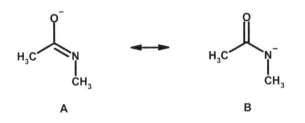

A   B

6)   In the end, each resonance structure should have the same overall charge and total number of electrons (bonds + lone pairs) as when you started.

7)   Remember the Golden Rule of Resonance: Never break a single bond.

Take Home Message: **Resonance is like telemarketers. They are never going to go away, so you need to learn them well.**

#12- You Spin Me Right Round Baby, Right Round, Like a Record Baby.

3-D visualization is one of the hardest aspects of organic chemistry for new students to learn and understand. One of the great aides in this area is a molecular modeling set, which will allow you to hold a representation of a tetrahedral carbon in your hands and rotate it yourself.

Along those lines, assigning R/S configuration to a chiral carbon can be made more difficult if the lowest priority group is not in the rear. For example:

It not completely obvious that this is the R stereoisomer from this view. There are a couple of things that we can do to this molecule to make the configuration more apparent. First, we could think of this as a Newman projection instead of the standard tetrahedral carbon.

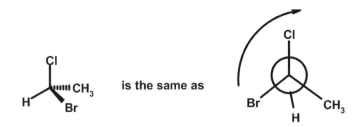

The other (easier) way to do this is to just rotate the tetrahedral carbon. The way to do this is to freeze one of the groups and rotate the other three. As long as they are rotated and not switched, (they can be rotated in either direction, clockwise or counterclockwise) it will be the same molecule.

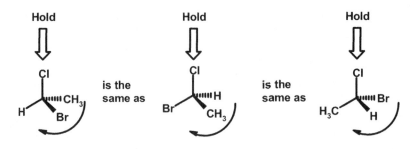

Any of the four groups can be held in place while the other three are rotated. Once rotated to a point where the lowest priority group is in the back, it becomes much easier to assign R/S configuration. Further, this method is not just confined to small molecules, as shown below.

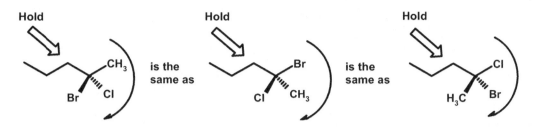

In the example above, CH_3 is the lowest priority group. By keeping the alkyl chain stationary, we can rotate the other groups clockwise in order to place the methyl group in back. Now it is more apparent that it is in the R configuration.

Take Home Message: Don't get angry. Just rotate your molecule to place the lowest priority group in back.

Bonus: Play with your molecular modeling kits

Many of your professors will require you to purchase a molecular modeling kit. Whether your kit is an old wooden one, the newer plastic ones, or one of the ultra high-tech magnetic ones, they are all helpful in learning this class. There are a number of benefits to using them routinely:

1) You will get a feel for what the geometry of the different structures is like.
2) You will become more comfortable with how many bonds a certain atom might have at any one time
3) You will quickly see that some molecules don't like to get together because the bonds are too strained.
4) These can be a big help in determining R/S configuration.

All in all, we huge fans of modeling kits and strongly suggest that every first semester organic chemistry student have one at their disposal.

#13- Size Matters: Resonance between equivalent atoms means equal bond lengths.

Once most students hear this tip, it makes perfect sense to them, but it isn't one that you might think about on your own. Take a look at the structure below, and ask yourself: are the two N-O bonds the same length?

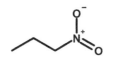

Since freshman chemistry, we have been told that double bonds between two atoms are shorter than a single bond between the same two atoms. Hence, the N-O double bond should be shorter than the N-O single bond. But let's look at some resonance forms:

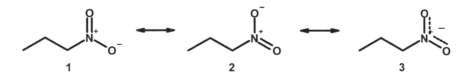

Here, we can more clearly see that the nitro group is moving between the three resonance structures. Structure 3, where the charge is spread evenly between the two oxygens is a valid structure and shows that the bond two oxygen atoms in the molecule are equivalent and have the same bond length (124 pm).

We care about this principle when it can be applied to more complex organic molecules where it is not obvious that the atoms are equivalent. Take the cyclopentadiene anion:

At first glance, this appears to have three different carbon atoms. However, once you start looking at resonance structures, you can see that the anion can be moved to any of the carbons in the ring. This makes them all equivalent, via resonance. This is confirmed through analytical studies which show that all C-C bonds are approximately 137pm long. Additionally, as this fits Huckel's rule of 4N+2, the molecule is also aromatic.

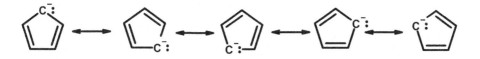

Take Home Message: If you see symmetry or aromaticity, think equivalent atoms

#14- Good for nothing alkanes. Lousy molecules.

Let's face facts: alkanes don't have many uses. In real life, they are non-polar solvents, larger-chained ones can be used to make wax, and they are good for burning/they make good fuels. As far as organic chemistry goes, alkanes are very boring. They have only three uses in your class:

5) Solvents: Alkanes are very good non-polar solvents. See tip #15 for more information on this.

6) Halogenation: Alkanes can be reacted with bromine or chlorine under free-radical conditions to obtain an alkyl halide. See tip #43 for more information on this.

7) Combustion: This is the fuel part. Complete combustion is an alkane reaction with oxygen to obtain CO_2 and water, as shown in the following example: $C_3H_8 + 5O_2 \rightarrow 3CO_2 + 4H_2O + Heat$

There is only a small possibility that you will see a combustion reaction on one of your exams. Therefore, the only real uses for alkanes in your undergraduate organic chemistry class will be as a solvent or in a halogenation reaction. Hence, if you have an alkane in one of your exam reactions, it should be very simple to determine its role. If there is a halogen (usually either Br_2 or Cl_2), then beware of a halogenation reaction. Otherwise, it is most likely safe to say that if you have an alkane in your reaction, it is a solvent and does not participate as a reactant.

Take Home Message: If your reaction does not have Cl_2 or Br_2 somewhere, your alkane is most likely just a solvent.

#15- In organic chemistry, "likes dissolve likes".

This one goes back to your days in freshman general chemistry and it is a simple one, but important to organic chemistry. In our world, likes dissolve likes. This means that polar molecules are easily dissolved in polar solvents and non-polar solvents are easily dissolved in non-polar solvents.

Now that we have comes to terms with this fact, the next question is: what constitutes polar molecules and solvents?

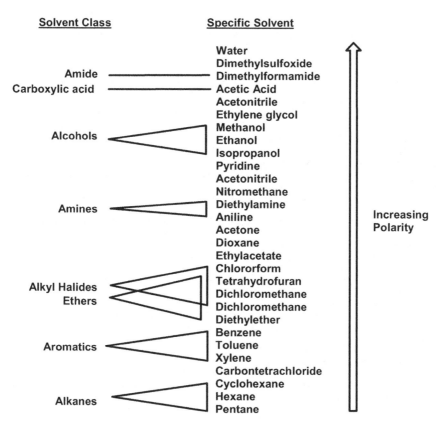

The chart above very effectively displays where different solvents rank with respect to their polarity. What should you take away from all of this?

1) There is no need to memorize the chart. Just generically learn which groups are more polar than others.
2) Alkanes and aromatics are very non-polar.
3) Water and alcohols are the main polar solvents.
4) **POLAR MOLECULES ARE EASILY DISSOLVED IN POLAR SOLVENTS. NON-POLAR MOLECULES ARE EASILY DISSOLVED IN NON-POLAR SOLVENTS.**

The polarity of molecules to be dissolved can be estimated using the chart above. Further, within classes of molecules, polarity can also be estimated.

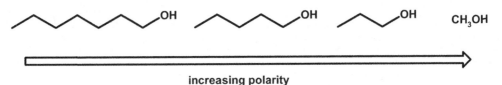

increasing polarity

Above, we have four alcohols. As we move from left to right, and the non-polar alkyl substituent is shortened, the overall polarity of the molecule increases. Therefore, it can be concluded that adding non-polar substituents to a molecule will *decrease* its overall polarity.

Further, we can continue to differentiate between polar solvents by designating them as polar protic or polar aprotic. This is important when working S_N1/S_N2 problems. (See tip #22)

Some Polar Aprotic Solvents (Favor S_N2 reactions)

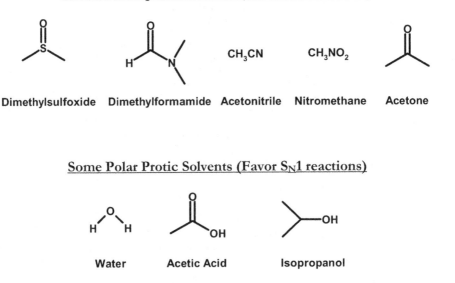

Dimethylsulfoxide   Dimethylformamide Acetonitrile Nitromethane   Acetone

Some Polar Protic Solvents (Favor S_N1 reactions)

Water   Acetic Acid   Isopropanol

Take Home Message: Polar solvents dissolve polar molecules. Non-polar solvents dissolve non-polar molecules.

#16-Beware of the bad acid trip: Meet your strong acids.

A problem we see students constantly running into is that they do not readily recognize strong acids. This is a terrible mistake and should never happen; you will need to quickly recognize strong acids and understand which atom they will be protonating. As far as strong acids go, you should immediately identify the Magnificent Seven:

STRONG ACIDS: HI, HCl, HBr, HNO_3, H_2SO_4, $HClO_3$, H_3PO_4

Just think that that it's the three hydrogen halides and the four acids that have N, S, Cl, and P as central atoms. Many students remember the other four acids with the mnemonic, something you would never say to a girlfriend: "**N**ever **S**ay '**P**lease **Cl**ean' ".

Once you have recognized that you have a strong acid present in your reaction, it is necessary to determine what it is protonating. Remember that H^+ is electron-deficient (Lewis acidic) and will look for an electron-rich (Lewis basic) atom to protonate. This could be a nitrogen atom, such as an amine, or an oxygen atom in a carbonyl or an alcohol. Most often, the atom being protonated will have a lone pair somewhere on it.

Take Home Message: Know the strong acids, the mnemonic "Never Say Please Clean" might help

#17- Meet your strong nucleophiles.

This is important throughout organic chemistry, but will be especially important when trying to determine the products of elimination and substitution reactions. There are generally three trends to remember when discussing how nucleophilic a reactant is:

1) Size - Generally, the more linear and/or smaller the nucleophile, the *more* nucleophilic it will be. This is because it can react at more sites and will not be sterically hindered if it is smaller or linear.

2) Electronegativity- The more electronegative an atom is, the *less* nucleophilic it will be. This is because more electronegative atoms want to hold electron density closer, and therefore will be less likely to let that electron density participate in a reaction. We see this in calculations and experiments that show nucleophilicity decreases as you get closer to fluorine on the periodic table ($C > N > O > F$)

3) Polarizability- The more polarizable an atom is, the *more* nucleophilic it will be. Polarizability is defined as the ability to distort the electron cloud of an atom, which allows it interact with a reaction site more easily. Generally, polarizability increases as you travel down a column of the periodic table ($I > Br > Cl > F$)

Below is a table of relative nucleophilic strength. This is relative because nucleophilic strength is also dependent on other factors in the reaction, such as solvent.

VERY Good nucleophiles	HS^-, I^-, RS^-
Good nucleophiles	Br^-, HO^-, RO^-, CN^-, N_3^-
Fair nucleophiles	NH_3, Cl^-, F^-, RCO_2^-
Weak nucleophiles	H_2O, ROH
VERY weak nucleophiles	RCO_2H

As shown above, as a general rule, the anion of a reactant will be a better nucleophile than the neutral form. (i.e. RCO_2^- is a better nucleophile than RCO_2H)

#18- They have worn out their welcome--Know your leaving groups.

Have you ever met that person seems to always be saying "screw you guys, I'm going home." That is a leaving group. They don't even need a good excuse to depart, they just do sometimes. So, what is the one overriding similarity between that guy and a good leaving group? Simple, they both enjoy being negative.

Leaving groups are atoms and molecular fragments that can easily support a negative charge through resonance, electronegativity, or polarizability. Below is a table of the most common leaving groups and their "nicknames".

Formal name	Structure	Nickname
Iodide	I⁻	None
Chloride	Cl⁻	None
Bromide	Br⁻	None
Methanesulfonate	$H_3C-S(=O)_2-O^-$	"Mesylate" or "-OMs"
Toluenesulfonate	$H_3C-C_6H_4-S(=O)_2-O^-$	"Tosylate" or "-OTs"
Trifluoromethanesulfonate	$F_3C-S(=O)_2-O^-$	"Triflate" or "-OTf"

Halides are generally known to be good leaving groups, but lesser known are the sulfonate esters. Using resonance, these special esters can distribute the negative charge over a number of atoms, thus stabilizing it and making it labile. In fact, the sulfonate esters are between 10^3 and 10^8 times more labile leaving groups than the halides.

Also, another type of good leaving group is the positively charged portion of a molecule. The classic example of this is when a hydroxyl (poor leaving group), becomes positively charged (aka water, a very good leaving group). See below for some examples where leaving groups are used in substitution or elimination reactions.

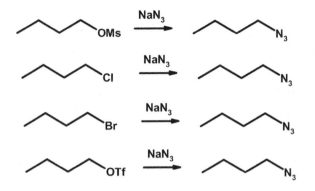

Take Home Message: Good leaving groups are halides, sulfonate esters, and water.

#19- If you don't start with chirality, you can't end with it.

This might be one of the finer points of the course, but it is an important one nonetheless. Some professors are sticklers about this point and frankly we think they should be. The rule (and make no mistake, it is a rule) is **if you don't start with chirality somewhere is your reaction, you cannot finish with it**. A chiral center cannot just be created magically. Almost always when you create a chiral center without any starting chirality, you have also created an equal amount of its enantiomer, making it a racemic mixture.

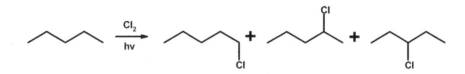

In the above reaction, the free radical halogenation of pentane produces three different products, none of which possess chirality. 1-chloropentane is not a chiral molecule for obvious reasons. 3-chloropentane is a meso compound and therefore also not chiral. 2-chloropentane contains a chiral center, but has its enantiomer present in equal amounts and therefore does not contribute to the overall chirality of the reaction mixture. This is because the chlorine radical has an equal chance of attacking from either face of the molecule, and therefore gives an equal mixture of each enantiomer. Because there is no chirality in the starting materials, the overall product mixture is also achiral.

The next logical question would be "how do you get chirality in your starting materials"? There are two methods for doing this: 1) chiral reactant, or 2) chiral catalyst. We will deal with chiral reactants, as chiral catalysts are a complex topic and will be discussed at a later time.

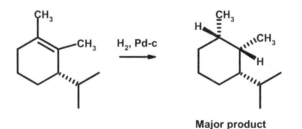

Major product

In the above example, a chiral center is already set in the molecule, so when the molecule attaches to the Pd-C catalyst, hydrogen will only be added from one face of the molecule, on the face opposite of the isopropyl group. This means that the chiral center established by hydrogenation will have the methyl groups pointing in the same direction as the isopropyl group.

#20- Markovnikov was a liar. A dirty, smelly, liar.

It is now time to step into the AceOrganicChem time machine. We travel back to Russia; the year is 1869. The czars are big, personal hygiene is not, and we find a young chemist named Vladimir Markovnikov working in his chemistry lab. Dr. Markovnikov has developed a "rule" for organic chemistry that says when adding a hydrogen halide (HX) to an alkene, the nucleophile (X⁻) will go to the more substituted carbon. The other way to think about this is to say that whichever carbon has more hydrogens, gets the H⁺ from HX. And thus, Markovnikov's rule was born. Since then, there have been a lot of ways to remember Markovnikov's rule:

- "Them that has, gets" (hydrogens)
- "The rich just get richer" (hydrogens)
- "The less substituted gets more lovin'" (hydrogens)

This was all good until the early 1930's, when chemist Morris Kharash began to further investigate Markovnikov's work and found numerous exceptions, or anti-Markovnikov reactions, that Markovnikov decide to ignore. Kharash concluded that Markovnikov's rule was basically a guess and should be renamed from "Markovnikov's rule" to "Markovnikov's speculation".

Nonetheless, we are less concerned with feuding dead chemists, and more concerned with how you use this on an exam. It is our opinion that you should not memorize which reactions are Markovnikov and which are anti-Markovnikov, but you should learn the theory and mechanism behind the reactions and be able to apply to any situation. The classic example here is HBr addition to propene, with and without peroxides.

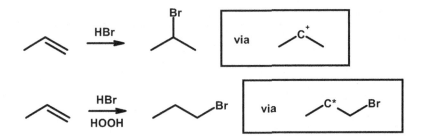

In the top reaction, we see Markovnikov addition of HBr across the double bond. This is because the first step of the reaction is the addition of H⁺ to the alkene, which when added to the carbon with more hydrogens, will provide a more stable secondary carbocation intermediate.

This is in contrast to the second reaction, which is called anti-Markovnikov addition. Here, the first step of the reaction is the addition of Br*. If added to the terminal carbon, the intermediate is the more stable secondary radical. Thus, this reaction provides 1-Bromopropane as the product, whereas the reaction without peroxides gives 2-Bromopropane as the product.

Some other examples of Markovnikov vs. anti-Markovnikov reactions are:

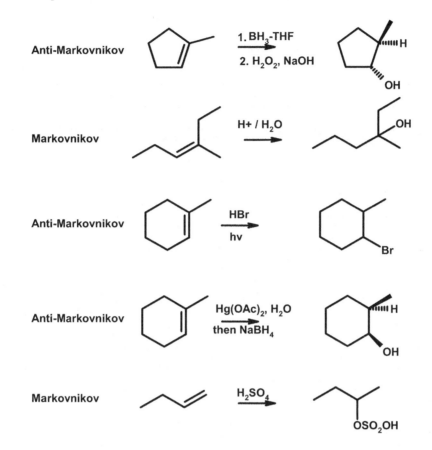

A better way to think of would be if addition of H+ is the first step, it probably goes through the Markovnikov addition because the more stable intermediate would be a more highly substituted carbocation. If the first step of the reaction is addition of a radical or addition of an electrophilic metal, then think anti-Markovnikov, because the more stable intermediate would be a more highly substituted carbon radical or an electrophilic metal on a more nucleophilic carbon. Either way, the key is to know which intermediate is more stable, then create a product from that.

<u>Take Home Message</u>: Know why a reaction occurs and you will never have to remember which reactions follow "Markovnikov's fabrication" and which ones don't.

#21- A different type of chemistry test. Functional Group Tests.

You should at least be aware of the different chemical tests for functional groups. Sometimes, these are given as part of an unknown molecule identification, or to supplement a spectral problem. If your professor is going to test you on them, then they will most likely be discussed in lecture. However, even if you are not going to be tested on them, it is worth at least one page in this book to talk about the main ones that can be used.

Test Name	Tests for	Positive test produces
2,4-DNP	Aldehydes and ketones	A yellow, orange-red, or red precipitate
Fehling's solution	Aldehydes	A red precipitate
Tollen's reagent	Aldehydes	A silver mirror
Iodoform test	Acetaldehyde and methyl ketones	A lemon yellow precipitate
Br$_2$	Alkenes	Colored solution turns clear

With the exception of the bromine test, all of the above tests are to chemically identify if your unknown is a carbonyl and which type it is. While these tests can tell you what class of molecule you are dealing with, it cannot distinguish between different reagents within that class. For example, a Tollen's test can tell you whether you have an aldehyde or not, but it cannot tell you whether or not you have formaldehyde or acetaldehyde.

Take Home Message: Chemical tests can tell you what class of molecule you have, but not specifically which compound it is.

#22- Is it E1, E2, S_N1, S_N2?

You are now forced to face the most common dilemma on an organic I exam. You have been provided a starting material and a nucleophile/base and are asked what the product/mechanism is. Are you looking at an E1, E2, S_N1, or S_N2 reaction? With some of these the answer can be very clear cut, like if you have a primary substrate and a non-basic nucleophile (S_N2). But what happens when it is not so clear cut?

Below is a chart that gives that conditions that help to promote each reaction.

	S_N1	E1	S_N2	E2
Reaction Mechanism	2-step with carbocation	2-step with carbocation	Concerted	Concerted
Strength of Nucleophile	Can be mediocre, must be non-basic	Can be mediocre, must be basic	Strong, non-basic, non-bulky	Strong and basic
Leaving Group Ability	Must be great	Must be great	Can be mediocre	Can be mediocre
Solvent	Polar protic	Polar protic	Polar aprotic	Polar aprotic
Product Stereochemistry	Racemic*	Alkene, bulky groups are trans to each other	Inversion of any chiral center	Alkene
Primary Substrate	No reaction	No reaction	Highly favored	Favored only with strong base
Secondary Substrate	Only with non-basic nucleophile or if carbocation is resonance stabilized	Only if carbocation is resonance stabilized (benzylic or allylic)	Need strong non-basic nucleophile, otherwise competes with E2	Favored with strong base
Tertiary Substrate	Favored with non-basic nucleophile	Competes with S_N1	No reaction	Can occur in strong base

*S_N1 reactions do not completely give racemic products. While it is true that the nucleophile has an equal chance of attacking either face of the carbocation formed, the leaving group will partially block one of those faces as it goes, which allows prevented complete racemization. For more information, search for "ion pairing" in your favorite search engine.

Before we proceed any further, a quick discussion of the four reaction mechanisms is in order. In is important to understand that in S_N1 and E1 reactions, the rate-limiting step is formation of the carbocation and that anything that facilitates that formation will also facilitate S_N1/E1. By the same token, the rate-limiting step in S_N2 and E2 reactions is pushing off the leaving group. Therefore, anything that helps the nucleophile displace the leaving group will speed up the S_N2/E2.

One of the tricks you can use is to write "SNLTS" (Remembered by the mnemonic "**S**aturday **N**ight **L**ive **T**otally **S**ucks") near each test question you have. "SNLTS" is to get you to remember the five factors of the reaction: Substrate, Nucleophile, Leaving Group, Temperature, and Solvent. Then under each factor, write the type of reaction that it favors.

For example, if your reaction is on a 1^0 substrate, write "E2/S$_N$2" under substrate. If your reaction has a strong base in it, write "E1/E2" under nucleophile.

Let's examine this a little closer. Here are the major factors in determining whether it is an E1, E2, S$_N$1, or S$_N$2:

1) <u>Substrate</u>: Primary substrates will **always** go S$_N$2 or E2. Tertiary substrates will *usually* go E1 or S$_N$1. Secondary substrates are in the middle and will mostly go S$_N$2 or E2. The exception to this is secondary substrates that are the resonance-stabilized carbocations, like benzylic and allylic carbocations.

2) <u>Nucleophile, strength and basicity:</u> Is it a strong nucleophile? (See tip #17 for strong nucleophiles). Strong nucleophiles will promote S$_N$2 because the leaving group has to be forced off. In S$_N$1, the nucleophile is attacking a vacant carbocation and therefore does not have to be as strong. Is it also a strong base? Obviously, bases will promote elimination, but remember that many nucleophiles are also strong bases.

3) <u>Leaving Group</u>: Mediocre leaving groups, like Cl, promote S$_N$2 and E2 because they are, by definition, forced off of the molecule. Great leaving groups, such as I, will promote E1 or S$_N$1. This is because removal of that leaving group and formation of the cation is the rate-limiting step.

4) <u>Temperature</u>: Elimination is generally favored at higher temperatures ($>50^0$C) while substitution is favored at lower temperatures.

5) <u>Solvent</u>: Is the solvent protic or aprotic? Protic solvents always promote S$_N$1/E1, because they stabilize the cation intermediate better. Polar aprotic solvents promote S$_N$2/E2 because they do not hydrogen bond and cannot interfere as much with the attack of the nucleophile.

In our opinion, one of the best clues for determining whether you have an S$_N$1/E1 or S$_N$2/E2 is the solvent. Professors are very careful to choose the solvent for test questions and will most often follow the rule of protic solvent for S$_N$1/E1 and aprotic for S$_N$2/E2.

One point that must also be stressed is that eliminations are always competing with substitutions and vice-versa. As they say on one of my favorite cable TV networks, "you can't stop it, you can only hope to contain it." On paper and on exams, we can point to some reactions and say that there is only one product, but in the lab you will almost always have a mixture of products in these reactions.

<u>Take Home Message</u>: Examine the five major factors to see if you have an S$_N$1, S$_N$2, E1 or E2 reaction.

#23-Why you should not stay on the straight and narrow (if you are counting carbon chains).

For those of us who were Boy Scouts, Girl Scouts, Brownies, or just spent a lot of time in detention, you have had an authority figure tell you about the virtues of staying on the "straight and narrow". Organic chemistry is one of those places where that advice should be avoided—if you are talking about nomenclature and carbon chains you should avoid limiting yourself to just looking at the straight.

We have yet to see an organic I exam where the professor did not dust off this ageless trick. Really, not a single one.

Q. Name this compound

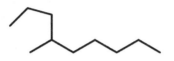

The organic chemistry novice (or the student who stayed up way too late watching infomercials and eating burritos) would ***INCORRECTLY*** name is compound 2-propyl heptane, when the correct name is **4-methyl nonane**. The mistake made is in assuming that the carbon backbone is in a straight line across your page. Time and time again, lazy professors will attempt to catch even lazier students in this trap. It is essential to know that you are naming the longest carbon chain, not necessarily the straightest. We suggest that you avoid this problem by counting the carbons, finding the longest contiguous chain of them, and placing a circle around it (our "circle" is in a large pentagon). Then, finish the job by placing boxes around any substituents. Now, our molecule looks like this.

Q. Name this coumpound

The other pitfall to avoid in the nomenclature game is assigning the wrong numbers to substituents. Substituent groups are always given the lowest possible starting number, regardless others that may be present. For example, the molecule below would be 2, 5, 5-trimethylheptane and *NOT* 3, 3, 6-trimethylheptane. Even though the second one might look ok, the first one is the correct name.

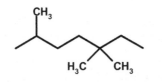

Correct: 2,5,5-trimethylheptane
Incorrect: 3,3,6-triemethylheptane

Take Home Message: Go for the longest carbon chain, not necessarily the straightest.

#24-Arrows will point the way.

While it may seem like a minor point, different arrows in organic chemistry mean different things:

⟶	Straight single-headed arrow Shows an irreversible reaction
⇌	Two arrows Shows a reaction equilibrium
↔	Double-headed arrow Shows resonance
⤵	Curved full arrow Shows flow of electrons
⤸	Curved half arrow Shows flow of single electron (radical)
⟹	Large single arrow Used for retrosynthesis

The most important thing to understand with these arrows is the electron flow with curved arrows. This will be an integral part of any mechanism that you will draw. As a strict rule, electron flow is **_always_** drawn showing where the electrons are _moving to_. This means that electrons will flow from negative to positive, from nucleophile to electrophile, or from base to acid.

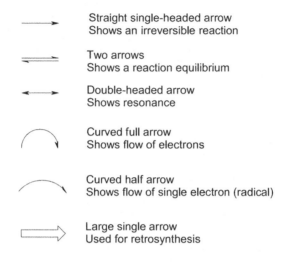

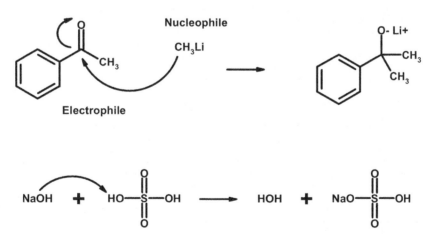

#25- There are only 3 different mechanisms to know. That's it, just 3 (kinda).

Sometimes people make things far more difficult than they need to be. Relationships, computers, and saving money on your car insurance are just a few examples of this. A fourth example is organic chemistry mechanisms. For the most part, in its most basic form, you will mainly only be exposed to three different mechanisms in undergraduate organic chemistry. It's hard to believe, but that is it. Of course, those three major categories of mechanisms can be further broken down, but once it is recognized that there are only three different mechanisms, organic chemistry gets much easier.

Type 1: Substitution reactions- In a substitution reaction, you are replacing one group with another. This can be done by reacting your substrate with a nucleophile, electrophile, or radical.

1a. Nucleophilic Substitution Reactions

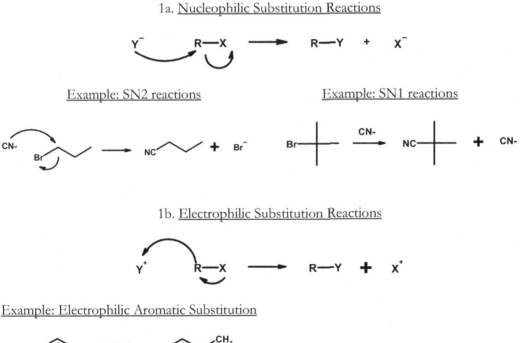

Example: SN2 reactions   Example: SN1 reactions

1b. Electrophilic Substitution Reactions

Example: Electrophilic Aromatic Substitution

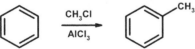

1c. <u>Radical Substitution</u>: note the single-headed arrows

<u>Example: Radical Propagation</u>

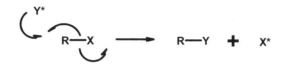

Type 2: Addition to Double and Triple Bonds- Addition to double and triple bonds occurs in several fashions. This can either be via electrophilic addition, radical addition, or a cycloaddition.

2a. <u>Electrophilic Addition</u>

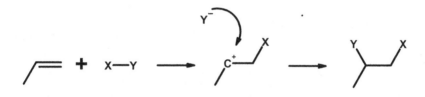

<u>Example: Acid halide addition</u>

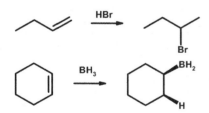

2b. Cyclic Reactions

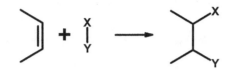

Example: Diels-Alder

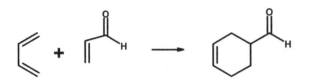

2c: Radical Addition

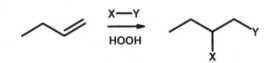

Example: Acid halide addition with radicals

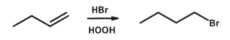

Type 3: Beta-Elimination- These are reactions that require a base and proceed via two types of mechanism, E1 or E2 (each of which you should be very familiar with).

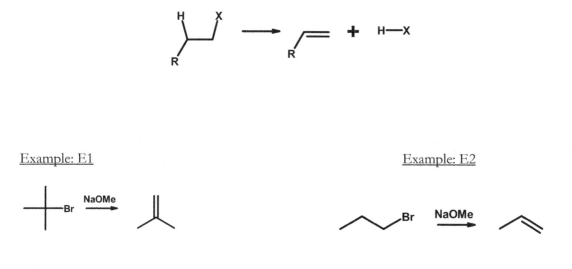

Example: E1   Example: E2

Take Home Message: There are very few actual mechanisms to learn. The key is to learn which mechanism type applies to your specific reaction.

#26- "They have the internet on computers now??"

The internet is a wonderful thing and a great source for organic chemistry help. Below are some very helpful organic chemistry sites (in no specific order) that will get you the grade you want.

A spectral database, with ^{1}H, ^{13}C, IR and MS of many organic compounds	Functional Group Chemistry: An e-book
http://riodb01.ibase.aist.go.jp/sdbs/cgi-bin/direct_frame_top.cgi	http://books.google.com/books?id=Kkgbi0uXWFgC&printsec=frontcover&source=gbs_summary_r&cad=0
Comprehensive online text	Comprehensive online text
http://www.chem.ucalgary.ca/courses/351/Carey5th/Carey.html	http://www2.chemistry.msu.edu/faculty/reusch/VirtTxtJml/intro1a.htm
Organic Reaction Mechanisms: An advanced e-book	An organic chemistry glossary
http://books.google.com/books?id=w9lTxIaPCQIC&printsec=frontcover	http://www.chemhelper.com/glossary.html
Organic chemistry help board	Name reactions
http://www.chemicalforums.com/index.php?PHPSESSID=811738b7ebc34ed2c443ef09253b8abb&board=3.0	http://www.organic-chemistry.org/namedreactions/

Organic Chemistry Help	Synthesis database
http://www.aceorganicchem.com	http://www.orgsyn.org
Organic compounds constants database	**MIT chemistry course webcasts**
http://www.colby.edu/chemistry/cmp/cmp.html	https://ocw.mit.edu/courses/find-by-topic/#cat=science&subcat=chemistry&spec=organicchemistry

Want to Ace Organic Chemistry?
Even if you currently have an F?

Get the Ultimate Vault of **Proven Study Tools** to Ace Your Organic Chem Exams
You get a 50% discount just for buying this book.

 "I went from a D to an A. Thank you so much!".

 "Easily explained some difficult concepts. Visual aids REALLY helped. Easy to understand and follow presenter."

"Your Organic I videos helped me so much for the first semester. You guys are doing a great thing!!"

With your enrollment you get:

- Videos, flashcards, eBooks, mini-movies, practice exams, and MUCH more proven to get you results.
- Weekly emails from your personal Sherpa, telling you what to study with links to find it, save you study time.
- Study plan with links to the material, based on the grade you want.
- 24/7 access anytime, anywhere on any device, to study on your time.
- 24/7 support to ensure your success
- Material is continually created to give you even more to help.

There is no risk to try it:

- Reason 1: Cancel anytime 24 hours, no questions asked.
- Reason 2: You get to keep the eBooks and study guides if you do cancel. The books go for $30 on Amazon.
- Reason 3: We take care of our students. We succeed when you succeed.

No Risk Discount Enrollment at
www.AceOrganicChem.com/special.html

<u>Chapter 3</u>: Organic Chemistry Strategy

#27- Go to the hardware store and get a chemistry toolbox.

Realize it or not, you are creating a toolbox of reactions that you can use your exams. Some of these reactions are trivial, like a 40-tooth saw blade, and you might never use them. Other reactions/techniques are essential, like a hammer or screwdriver, and you will use them all of the time. And just like in your real toolbox (does anyone but me keep a toolbox anymore?) you want to sort your tools by usefulness, keeping the best ones on top and placing the rest out of the way. Here, we are going to help you sort your toolbox a little bit.

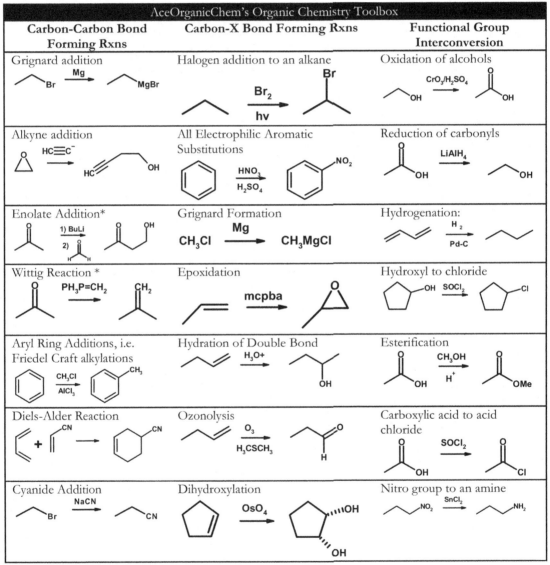

AceOrganicChem's Organic Chemistry Toolbox		
Carbon-Carbon Bond Forming Rxns	**Carbon-X Bond Forming Rxns**	**Functional Group Interconversion**
Grignard addition	Halogen addition to an alkane	Oxidation of alcohols
Alkyne addition	All Electrophilic Aromatic Substitutions	Reduction of carbonyls
Enolate Addition*	Grignard Formation	Hydrogenation:
Wittig Reaction *	Epoxidation	Hydroxyl to chloride
Aryl Ring Additions, i.e. Friedel Craft alkylations	Hydration of Double Bond	Esterification
Diels-Alder Reaction	Ozonolysis	Carboxylic acid to acid chloride
Cyanide Addition	Dihydroxylation	Nitro group to an amine

*2nd **Semester reactions**

We have now gone to the largest hardware store in the country and started to create your organic chemistry toolbox. We have divided your new toolbox up into three different sections: carbon-carbon bond formation, carbon-X bond formation, and functional group interconversion. Further, the most important of these reactions are placed at the top of each column. While each of the sections of the toolbox is equally important, it is necessary to segregate them in order to be able to understand which reaction to use in which circumstances.

1) Carbon-Carbon Bond Forming Reactions: These are the most important of the reactions. Most of your synthesis problems will force you to use a starting material that has fewer carbons than your final product. You will need to use one (or more) of these reactions to get to the right number of carbon atoms in your final product.
2) Carbon-X Bond Forming Reactions: In addition to having the right number of carbon atoms, you must also have the right functional groups on your molecule in a synthesis problem. These are the handiest reactions to use to get those functional groups on to your molecule.
3) Functional Group Interconversion: This is usually the simplest step. It is at this point that you already a necessary functionality on your molecule, and just need to change it around a little.

A very helpful exercise is to sit down and do this for yourself. Each professor is different and will emphasize different reactions, so your toolbox may have some reactions that ours does not. Take some time and think through which reactions are the most useful for your class and get yourself a good toolbox to ace those synthesis problems.

Take Home Message: Make a toolbox of the reactions which will be most helpful for your specific class.

#28- It Takes Alkynes to Make the World Go Around (and to react with epoxides)

Epoxides are electrophilic three-membered rings with an oxygen inside. Alkynes, once deprotonated, are great nucleophiles and are very good at forming carbon-carbon bonds. A generic example of this is shown below:

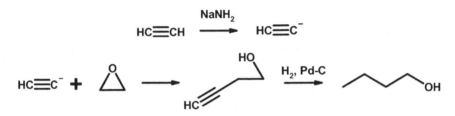

In the above reaction sequence, acetylene is deprotonated with a strong base. In this case, sodium amide is used. [Warning: NaOH and alkoxide bases are not strong enough to deprotonate acetylene]. Next, the acetylene anion is reacted with the epoxide. The acetylene adds to the epoxide and opens the ring to form an alcohol. In the final step, the alkyne is reduced to the alkane by hydrogenation to give the final product, which is the alcohol shown. Thus, we have shown a generic method for adding a two-carbon anion to a two-carbon epoxide to obtain a four-carbon alcohol.

While the reaction shown above is relatively simple, more complex reactions can be used to obtain more complex products. For example:

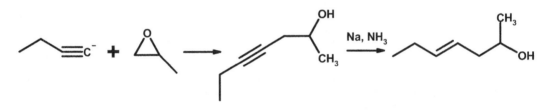

In the case above, we reacted a longer chained alkyne with a substituted epoxide. The general rule for epoxide opening is that under basic conditions [alkynes are basic nucleophiles] the nucleophile will attack at the _less_ substituted carbon, as the nucleophile will go to the less hindered side. In this reaction, after the alkyne attacks, we reduce the resulting compound from the alkyne to the trans-substituted alkene with sodium and ammonia. Thus, this carbon-carbon bond forming reaction is good for creating substituted alcohols and for incorporating specific double bonds into our target molecule.

Take Home Message: Alkynes are a great way to attack an epoxide and synthesize an alcohol with a double bond in it. If you don't need the double bond, then just use the Grignard reagent instead of the alkyne.

#29- Four organometallics to rule them all

Grignard reagents are a wonderful invention. Like showers, deodorant, and hot dogs, they just make life so pleasant. The problem is that Grignard reagents are not the best reagent for every reaction you may run across. What if you want to do a substitution reaction on a halide? That is not a job for a Grignard reagent. The big secret in the organic chemistry world is that there are other organometallic reagents that will do just fine that are just as easy to form. Here, we will discuss the four types of organometallic reagents as well their best uses.

Grignard (RMgX):
$$CH_3Cl \xrightarrow{Mg} CH_3MgCl$$

OrganoLithium (RLi):
$$CH_3Cl \xrightarrow{Li} CH_3Li$$

Organocuprate (R$_2$CuLi):
$$2\ CH_3Li \xrightarrow{CuCl} [(CH_3)_2Cu]\text{- } Li^+$$

Organozinc (RZnX)
$$CH_3Cl \xrightarrow{Zn} CH_3ZnCl$$

	RMgX	RLi	R$_2$CuLi	RZnX
Basic?	Yes	Yes	No	No
S$_N$ rxn with RX?	No	Yes, with some elimination by-product	Yes	Yes, on dihalides to form C-C bonds
Reacts with C=O?	Yes	Yes	No	No
Reacts with esters?	Yes	Yes	No	No
1,4 addition to α,β unsaturated C=O	Will do 1,2 addition first, then 1,4	Will do 1,2 addition first, then 1,4	Yes	No
Can form carbene	No	No	No	Yes

I)   Grignard Reagents: These are the good old standby. Like an old t-shirt, they are comfortable, so you know and love them. Formed by mixing an alkyl halide with Mg, they will react with almost any electrophile. The most common use in for attacking at a carbonyl carbon. They are basic, and relatively stable, but will not do substitution reactions on alkyl halides. Sample reactions are shown below:

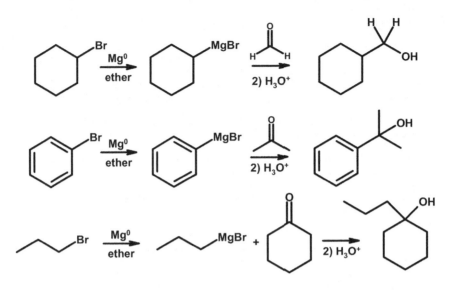

II)   Organolithium reagents: These are similar to Grignard reagents, but have one
        major benefit: they can displace a halide in an S_N2 reaction, where as a Grignard
        cannot. Formed by reacting an alkyl halide with lithium metal, these nucleophiles
        can also be used as just a strong base it needed. This means that if they are to be
        used as a nucleophile in an S_N2 reaction, there will also be some E_2 product
        observed, due to the high degree of basicity.

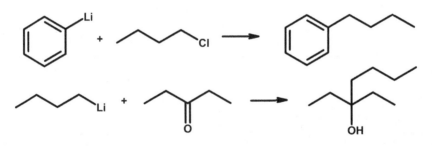

III)   Organocuprate reagents: While not the first choice of many organic chemistry
        students, organocuprates can be very useful. Formed by reacting an alkyl lithium
        with copper chloride, these reagents can perform 1,4 additions to unsaturated
        ketones and displace halides in an S_N2 fashion. The benefit to the S_N2 reaction is
        that the organocuprates are not basic, and therefore will not give any elimination
        by-products.

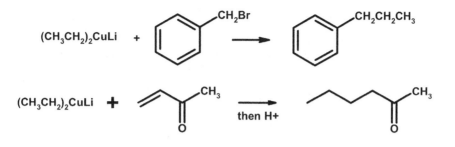

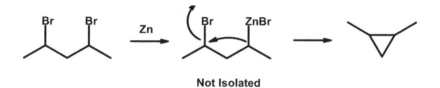

IV)   Organozinc reagents: Organozincs are like that weird second cousin we all have. Nobody really knows about them, and whenever they show up, strange things happen. Organozinc formation is similar to Grignard formation, where zinc is inserted in between the carbon-halogen bond. However, organozincs really only have one function, which is to couple of the carbons of dihalides.   For example, if the starting material is a 1,3 dihalide, then the product will be a cyclopropane.

Take Home Message: There are more organometallic reagents out there than just Grignards. Be aware of the others so you can use the right one for the right job.

#30- Diels-Alder Reaction-Part 1: An introduction

Would it have the same ring if it were called the "Kurt-Otto" reaction? Probably not. This reaction is one that many professors like to emphasize because it is one of the only cycloaddition reactions that students are shown in undergraduate organic chemistry. The reaction in its most basic form is very simple:

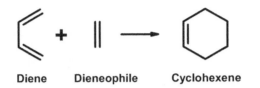

Diene   Dieneophile   Cyclohexene

In this reaction, a diene (by definition having two double bonds in conjugation) and a dieneophile (something that "loves" a diene) are reacted to give a cyclohexene ring as the product, which can be hydrogenated to give a cyclohexane ring. The reaction is considered a [4+2] cycloaddition, with the generally accepted mechanism shown below:

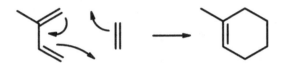

As a general rule, this reaction proceeds best with certain types of reactants. Specifically, the reaction proceeds best when the diene has electron donating groups attached to it and the dieneophile has electron withdrawing groups attached to it. Some examples of good dienes and dieneophiles are shown below:

Some Good Dienes

Some Good Dieneophiles

Some examples of Diels-Alder reactions are shown here:

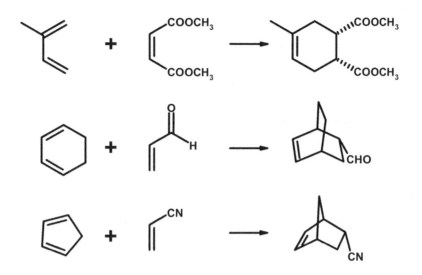

HINT: If your test question has a cyclohexane or cyclohexene ring as the product and you have to use starting materials with FOUR CARBONS or LESS, it almost certainly will involve a Diels-Alder reaction. [You can synthesize a cyclohexane ring by hydrogenating your product]. This is going to be one of your main carbon-carbon bond forming tools for your toolbox.

In Tips #54 and #55 we will further explore this reaction and go into some of the conditions that will make for a successful Diels-Alder reaction.

#31- Let's go retro: Retrosynthetic analysis

Ah, remember the 70's? Discos, a nice large 'Fro, and socks. Retrosynthesis is kinda like the 70's, except it actually has a purpose. The purpose of retrosynthesis is to give us a viable pathway to get from a small molecule to a large one. The theory behind it is that it is much easier to breakdown a large molecule to see which small molecule it could come from than the other way around. There are really two actions that can be taken in a retrosynthesis, as shown here:

Synthons:

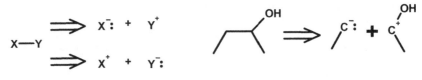

CAUTION: Synthons are not necessarily real

Functional Group Interconversion

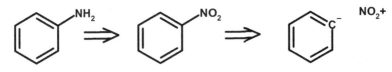

Synthons are the fragments that come from breaking a bond. The important point with synthons is to recognize that they are not necessarily real. They just represent the fragments that might be possible. For example:

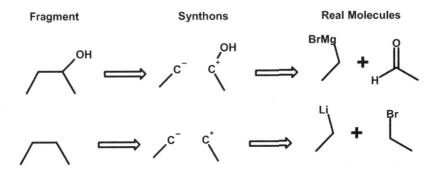

Here, we are cleaving strategic bonds. We assign a hypothetical "+" or "-" charge to each end of the broken bond to help us decide which real molecules our synthons should come from.

Once your major carbon-carbon bonds have been formed, it is time to change the functional groups in the groups we need for our final product. In this section, we will not talk a lot about FGI (functional group interconversion).

While entire books have been written just about retrosynthesis, here are some simple tips to help you complete your synthesis problems more effectively:

✓ DO: Work backwards to an "acceptable" starting material
✓ DO: Count your carbons so they don't get lost or forgotten
✓ DO: Locate functional groups, they will be the first & best place to look for disconnections
✓ DO: Place most reactive functional groups on product last
✓ DO: Examine more than one pathway (time permitting)
✓ DO: If you run into trouble, start interconverting functional groups looking for ideas.

✗ DON'T: Ignore other functional groups on the molecule.
✗ DON'T: Make up chemistry that "looks like it should work".
✗ DON'T: Use reactions that will give mixtures of product. (May lose points on exam)

Retro FAQ's:

Q. Where should I choose to disconnect?
A. More often than not, disconnections take place immediately adjacent to functional groups because they have electronegativity differences from carbon and therefore can do chemistry.

Q. How do I recognize a good disconnection?
A. A good disconnection will make the molecule visibly simpler.

Q. How do I decide which synthon carries the charge?
A. One trick is to use resonance. If you draw a synthon and the resonance structure looks similar to a real reactivity intermediate, you are most likely on the right track.

Q. Was disco ever cool?
A. No, they just didn't know any better back then.

#32- Overall EAS strategy

Electrophilic Aromatic Substitution (EAS) problems are like regular synthesis problems, but different in that they focus more on the order that groups are substituted on the benzene ring rather than on the transformations themselves. To be sure, chemistry done directly on an aromatic ring (like benzene) is very different than the chemistry on other organic molecules.

For example, you may be asked to synthesize the molecule below:

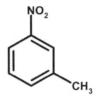

Again, the order of your reactions is important. Below, we show two methods to obtain nitrotoluene. The first method is the one that will give the correct product. Notice in the second example that switching the steps gets us a different product.

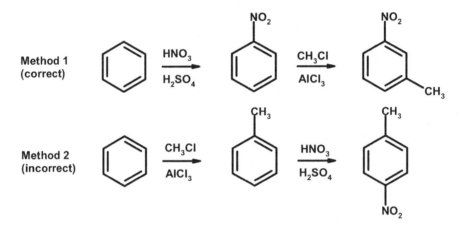

The reason for this is the "directing" effect of the first group placed on the ring. In general, groups that withdraw electron density from the aryl ring are deactivating groups and will tend to direct new groups to the meta position. Almost without exception, groups that donate electron density to the ring are activating groups and will cause incoming substituents to go to the ortho and/or para positions. Below is a chart of activating and deactivating groups:

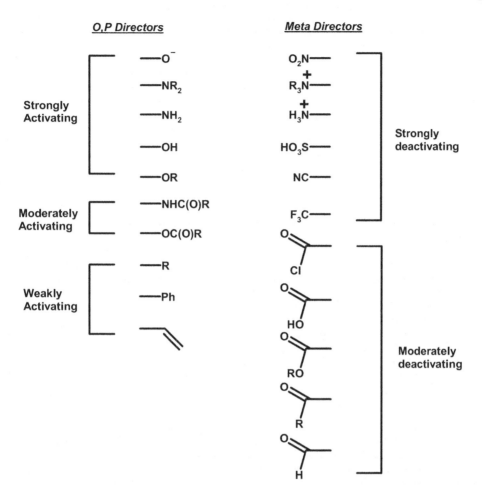

As a general guide, atoms that are attached directly to the ring that have free lone pairs and/or negative charges will be o,p directors. Conversely, meta directors are comprised of carbonyl groups, nitriles, and sulfates and positively charged atoms directly connected to the ring. The only exception to the activator/deactivator rule is halogens. Halogens are slightly deactivating, but act as o,p directors.

Now that we know the general rules of EAS, we can learn some of the tricks that we can use to synthesize more complex EAS products.

#33- EAS Strategy: Turn a meta into ortho/para and vice-versa.

A professor may give the following structure and require you to synthesize it from benzene.

The glaring problem here is that you have two o,p directors that are meta to each other. How the heck do you do that? There is an easy strategy to alleviate this issue, which involves placing a meta director on the ring first. This will send our second group into the meta position, then we convert the first group to achieve our end product. In the example above, we would first nitrate benzene, place our chlorine in the required meta position, then convert the nitro group to an amine to give the final product, as shown below.

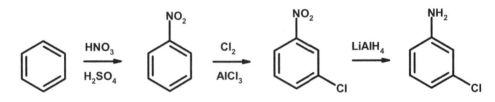

In EAS, there are a number of functional groups which are meta directors that can be converted to ortho/para directors later, and vice versa. Shown below, are a number of meta directors that can be converted into ortho/para directors.

Meta Director	Converted to o,p director	Conditions
NO_2	NH_2	$LiAlH_4$
CN	CH_2NH_2	$LiAlH_4$
COOH	CH_2OH	$LiAlH_4$
C(O)R	R	H_2, Pd-C
SO_3	OH	NaOH, HEAT!

Ortho/Para Director	Converted to meta director	Conditions
R	COOH	$KMnO_4$
NH_2	CN	HONO, then $Cu_2(CN)_2$
NH_2	NO_2	mcpba

When tackling any problem like this, the first thing that you should think to use is an amine if you need an o,p director, or a nitro group if you need a meta director. The advantage is that nitro groups can be converted to amines and amines can be converted into many other groups through diazo compounds. Diazos are highly versatile and will allow you to convert the amine to another group, or just remove it all together. Below is a summary of many of these reactions.

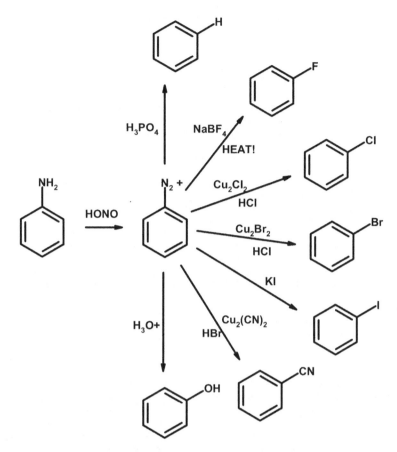

Your strategy for these types of problems should be to:

1. Identify the groups on your aryl ring. Which are o,p and which are meta?
2. Place the new substituents on the ring with the directors you have right now, if you do it in the right order.
3. If not, identify which of the current substituents/directors can be converted into a different type of director. (meta to ortho/para or vice versa)
4. If not, can you place an NH₂ on the ring somewhere, use it to direct, then remove it?

Take Home Message: If you are not sure how to get the groups on your aryl ring in the right place, consider changing the substituents on the ring to help you.

#34- EAS Strategy: Conversion of alkyl groups to carboxylic acids.

We see circumstances for this strategy arise more frequently than one might expect. The classic question that is seen on exams is shown below, where you are asked to synthesis the following molecule from benzene:

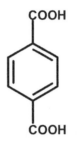

As in tip #33, the obvious issue is that COOH is a meta director and the two groups are para to each other. To solve this problem, you will use a strategy where you place ortho/para directors on the benzene ring first, then convert them to the meta director after you have them in the proper place on the ring. For this example, we would proceed as follows:

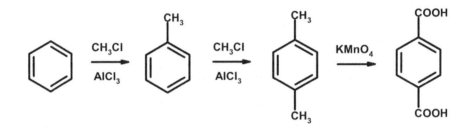

The first step is a Friedel-Craft alkylation of benzene to give toluene. This is then followed by another Friedel-Craft alkylation to give p-methyl toluene, also called p-xylene. The trick here is the reaction that turns alkyl groups to carboxylic acids by subjecting them to $KMnO_4$. This is referred to as a sidechain oxidation, and can be performed on a number of different alkyl groups. We now have converted both methyl groups into carboxylic acids and completed the synthesis.

Further examples of this reaction are shown below:

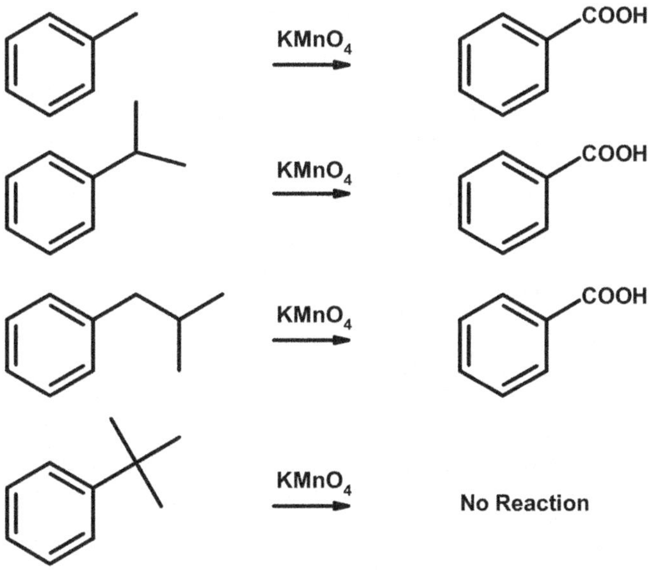

The reaction is a very handy one, and can be used on _almost_ any alkyl side chain. The key to the reaction is the benzyl protons on the alkyl side chain, which are necessary for the reaction to proceed. Without at least one benzylic proton the reaction will fail, as illustrated in the example of t-butyl benzene.

Take Home Message: Alkyl side chains can be converted into carboxylic acids. Keep this in mind if you see a problem where you have a carboxylic acid but can't find out how to place it on the ring.

#35- EAS strategy: -SO₃H and -X are our blocking groups

It's fourth down and we need to find the end zone. What is the best way to do that? Get yourself a fat blocker and run the ball in. In EAS, we have two fat blockers. At fullback, we have halogens blocking for us. At halfback, we have sulfonic acid. A typical problem where these groups can be useful is shown below:

Synthesize the following:

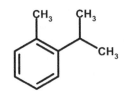

Care must be taken in this problem: both groups are o,p directors which means that once one alkyl group is placed on the chain, the natural tendency will be for the second to go to the para position, as shown below:

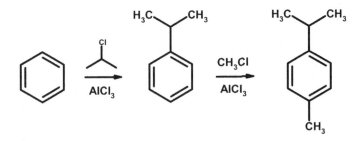

So how do we keep that second alkyl groups from going to the para position? Use a blocking group.

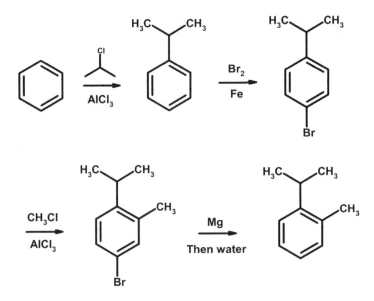

As shown above, after the first alkylation, we can use a halogenation reaction which will place a halogen group in the para position. Once the para position is successfully blocked, the second alkyl group will have no choice but to go to the ortho position. Using simple chemistry, we alkylated the ring and then halogenated, which is selective for the para position. Now, because the isopropyl group is still an o,p director, the next group will have to go to the ortho position. Friedel-Craft alkylation places the methyl group at the ortho position. The final step is to remove the halogen, which can be done with Mg and water.

SO_3H can also be used as a blocking group, if halogens are not appropriate, as shown below:

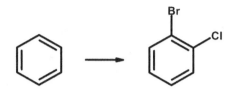

This synthesis can be accomplished by the following method:

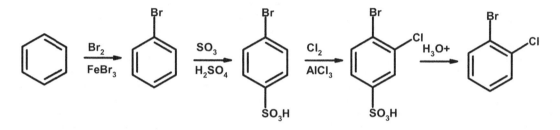

The strategy is the same, but using a different blocking group. Here, because we cannot use a halogen, the sulfate is used as the para-blocker. This group can be removed using aqueous acid in the final step.

Take Home Message: If you need an ortho substituted aryl ring, consider a blocking group.

#36- EAS Strategy: Long chain alkyl groups from Wolff-Kishner or reduction

Care must be taken when attempting to place a long alkyl chain on benzene. Consider the following problem:

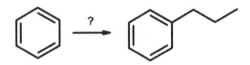

At first glance, this may appear to be a simple Friedel-Craft alkylation. However, upon further examination we remember that those alkylations proceed through a carbocation intermediate, which will rearrange to give a different product.

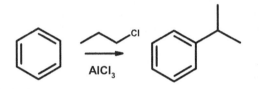

So how do we synthesize propylbenzene? When you need a long chain alkyl group on a benzene ring, place the chain on the ring via Friedel-Craft acylation, then remove the carbonyl in a later step.

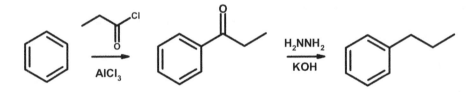

The carbonyl is removed using the Wolf-Kishner reaction, which then gives our final product. (See tip #58 for more information on the Wolff-Kishner Reduction)

Take Home Message: Alkyl chains that are three carbons or longer need to be place on the ring via acylation/carbonyl removal.

#37- EAS Strategy: Substituted toluenes came from toluene. Duh.

The acorn doesn't fall far from the tree. In our case, the substituted toluene doesn't fall far from toluene. Almost all substituted toluene molecules originally came from a substitution reaction on bromo-toluene.

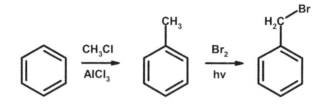

As shown above, the creation of bromo-toluene is started with a Friedel-Crafts alkylation of benzene to obtain toluene. This is then reacted with bromine under free radical conditions, which causes bromination at the alkyl side chain and not on the aromatic ring.

This compound is now useful because it has a leaving group on in the benzylic position, where it can undergo an S_N1 OR S_N2 reaction (See tip #40 for more information on benzylic leaving groups). Now that it is susceptible to nucleophilic substitution, we can look at a number of different reactions we create different molecules based on bromotoluene:

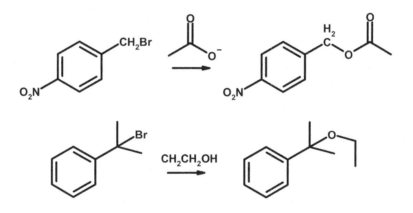

Thus, the retrosynthesis for almost every substituted toluene should look like this:

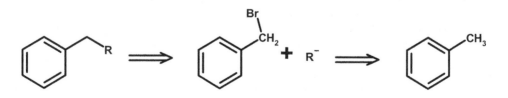

<u>Chapter 4</u>: Helpful Hints and Clues

#38: Want to impress your instructor? Start dropping names.

Picture this--It's Saturday night, you had a good time at the bars and are at a pretty good after-hours party now. You are talking to this great looking girl/guy and it is going well until you hear things like:

> "I was on the phone with Governor Hibbert last night, he is such a good friend, and he offered me an internship this summer..."
> "I know the whole basketball team. The star forward, Wiggum, stopped me on campus the other day to give me a kiss. He is so sweet"

Name dropping: One of the quickest ways to turn off that potential mate. Well, let me tell you where it is impressive. When you write a chemical reaction on one of your exams and put the name of that reaction next to the transformation (see example below), you have shown a greater mastery of the subject and are more likely to get that extra point if you somehow weren't perfect with the reaction.

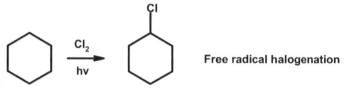

Free radical halogenation

Moreover, this can be essential if you are using a reaction on an exam that you were not taught in class. There are thousands of organic chemistry reactions that are perfectly valid chemical transformations that you were not shown in lecture. One of the main problems with using them on your exam is that your professor (or the TA grading the exam) might not immediately recognize this reaction and may erroneously dock you points. By placing the name of the reaction next to it, you very politely say to your TA: "Yes, I know what I am doing. Go look it up, dummy."

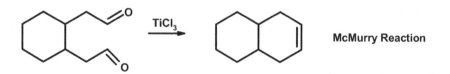

McMurry Reaction

The McMurry reaction is basically a "reverse ozonolysis" (see tip #62). It is a perfectly legitimate reaction and can be used in a number of situations, but is not generally taught in undergraduate classes. If your professor or TA is one that does not recognize this reaction, they may erroneously deduct points from your overall score. One way to help avoid this is to put the name of the reaction next to the transformation.

#39- Enantiomers are siblings and diastereomers are Cousins, which makes rotomers your dingy aunt.

The point we are trying to make with this tip is that the molecules are related when you compare them. You can't just walk up to someone on the street and know that they a sister; you have to know whose sister they are. Molecules are the same way. A molecule just isn't an enantiomer, you have to compare it to its mirror image to know it's the enantiomer. This is illustrated below:

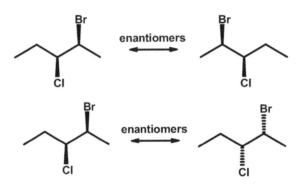

Here, we show two different ways you can turn a compound into its enantiomer. The first way is by drawing its mirror image, which is the definition of an enantiomer. You could keep the molecule's orientation the same and just reverse all of the stereocenters by turning the dashes to wedges and vice versa. Both are acceptable ways to determine this compound's enantiomer. These two methods can be applied to Fischer projections too.

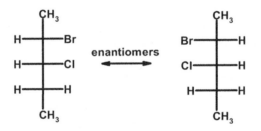

Determining diastereomers is just as easy. With diastereomers, the difference between the compounds shown would be that only one of the stereocenters in the molecule were different.

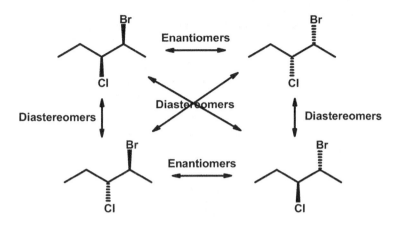

Keeping it systematic, as we have above, eliminates the problem of duplication of molecules. Moreover, we can even use a formula to determine the total number of possible diastereomers without actually drawing them all out

Total # of possible diastereomers = 2^N [where N is the number of chiral centers in the molecule]

In our example above, there are two chiral centers, so therefore we have 2^2 or 4 potential diastereomers. In the molecule below, we have far more than that.

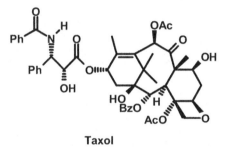

Taxol

Taxol is a potent anti-cancer drug derived from the Yew tree. Actually, changing one stereocenter at a time to determine the total number of diastereomers is not feasible, however it is feasible to count that there are 12 stereocenters. We can now use our formula determine that there are 2^{12} or 4096 possible diastereomers for Taxol.

Take Home Message: Use a link chart like above to systematically create all of the diastereomers for a molecule. Use the formula: Total # of possible diastereomers = 2^N to just determine the total number without drawing each one.

#40- Benzylic and allylic leaving groups are better than they look.

Overachievers can really be annoying sometimes. I mean, let's face it, who wants to watch that all of the time. This is your cousin that graduated from Yale to attend Princeton Medical School and helps homeless veterinarians in his spare time. Benzylic cations are the overachievers of organic chemistry. Upon first glance, the benzyl cation looks like an average primary cation. Since we know that primary cations are not stable, we would think that the benzylic cation would not exist.

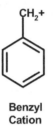

**Benzyl
Cation**

However, this is no ordinary cation. Because the positive charge is located one carbon off the aryl ring, this compound has resonance structures which make it much more stable than your average primary cation.

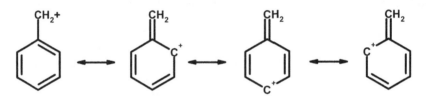

As shown above, the positive charge can be delocalized through the aryl ring to provide further stability. We are now spreading that charge over the entire aryl ring, instead of just one carbon atom. The greatly enhances the stability of the cation, giving it stability greater than a normal secondary cation. We can further enhance this stability be placing an electron donating group in the ortho or para positions.

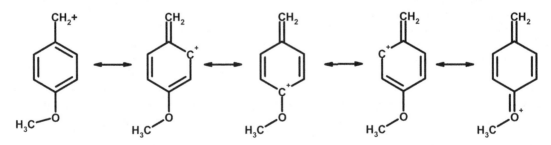

Now, the oxygen atom can also participate in the resonance structures, even further enhancing the stability of the cation.

For the same reasons, we can apply this logic to allyl cations too. However, in this case, the charge can only be spread over three atoms. While this is still more stable than an ordinary primary cation, it does not enhance it as much as the aryl ring example above.

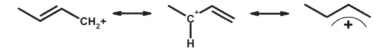

While allyl and benzylic cation are the overachievers of the cation family, vinyl cation are the underachievers. Because the charge is directly on the sp^2 carbon in a vinyl cation, there are no stable resonance structures that can be drawn. Further, the double bond actually serves to destabilize the cation to a small extent. So, although the cation below looks like a secondary cation, it actually has stability like it was primary.

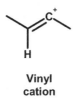

**Vinyl
cation**

We can now incorporate our newly found relatives into the overall family structure and compare their stabilities to each other.

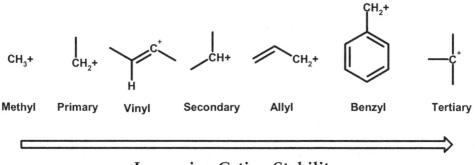

| Methyl | Primary | Vinyl | Secondary | Allyl | Benzyl | Tertiary |

Increasing Cation Stability

<u>Take Home Message</u>: Because of resonance, allyl and benzylic cations are more stable than they appear. Vinyl carbocations are less stable than they appear.

#41- Before you die, you see the [aromatic] ring

Back in the days of the early alchemists, compounds were named by their two most identifiable properties: smell and taste. While we would find this revolting today, the names for some of these compounds have stuck, which is where we get the name for aromatic compounds.

There are two criteria for a compound to be considered aromatic:

1) There must be 4N+2 pi electrons [where N is any integer, including 0]. Therefore, if a compound has 2, 6, 10 or 14 pi electrons, it has satisfied the first requirement of aromaticity.

2) The pi electrons must be conjugated in a contiguous ring. By this, we mean that the ring needs to continuously alternate between double and singles bonds throughout the entire ring. This is the tough one to determine sometimes, but we will help coach you through this.

Below are some of the easier examples of aromatic and non-aromatic compounds:

6 pi electrons, conjugated and contiguous therefore is AROMATIC

4 pi electrons, NOT contiguous therefore is NOT AROMATIC

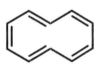
10 pi electrons, conjugated and contiguous therefore is AROMATIC

8 pi electrons, conjugated and contiguous therefore is NOT AROMATIC

While these examples are not too hard to navigate, there are other problems one may run into that are discussed below:

6 pi electrons, conjugated and contiguous therefore is AROMATIC

4 pi electrons, conjugated and contiguous therefore is NOT AROMATIC

2 pi electrons, conjugated and contiguous therefore is AROMATIC

4 pi electrons, conjugated and contiguous therefore is NOT AROMATIC

6 pi electrons, conjugated and contiguous therefore is AROMATIC

6 pi electrons, NOT contiguous therefore is NOT AROMATIC

The trick here is the positive or negative charge has been incorporated into the ring. These charges are like bridges and connect the double bonds to each other. Second, negative charges will supply two pi electrons, whereas positive will not supply any. So, if you have the right number of electrons, _counting the electrons from the charge_, and they are contiguous throughout the entire ring, then it will be aromatic.

The last trick in aromatic compound is nitrogen and oxygen-containing heterocyclic rings. The lone pairs of these heteroatoms may or may not participate in the aromaticity, depending on a couple of factors. The first factor is whether a bond can be strained to place the electrons in a position to bond and participate in aromaticity. In the case of pyridine, because of the double bond to nitrogen, the lone pair on nitrogen can never get into the proper geometry to participate in the pi electron system. This is not a problem, because they are not needed to achieve aromaticity. However, in the cases of pyrole and furan, the lone pairs are needed to achieve aromaticity. Therefore, the bond angles of the nitrogen and oxygen are strained a little bit in order to bend the lone pair into the correct geometry to participate as pi electrons. Interestingly enough, since there are two lone pairs on oxygen and only one is needed to get to the requisite aromatic magic number of 6, one lone pair participates in aromaticity while the other does not. Some say that aromatic molecules are like a snotty softball team in that once they get the right number of players, they don't accept any others.

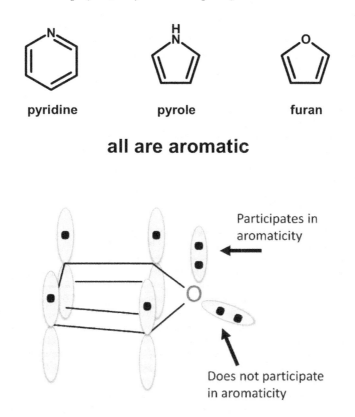

pyridine   **pyrole**   **furan**

all are aromatic

Take Home Message: There are two criteria for a molecule to be aromatic: 1) the right # of pi electrons, and 2) that they are in a contiguous and conjugated.

#42: "Eject! Eject! Eject!" Carbonyls with an ejectable group are added to twice.

These functional groups remind me of 1985 when Maverick flew through the jet wash and Goose and he had to eject from their F-14 Tomcat. What does this title mean? What we are trying to say is that carbonyls can be classified two different ways: ejectable or non-ejectable. What this means is that sometime when a carbonyl is attacked by a nucleophile the carbonyl will eject one of its substituents *before* it reduces the carbonyl to an alcohol. After the group has been ejected, then a second equivalent of nucleophile will reduce the carbonyl to alcohol.

In essence, this means if a carbonyl has an ejectable group on it, a nucleophile will add twice to that carbonyl. Some examples of ejectable and non-ejectable systems are below:

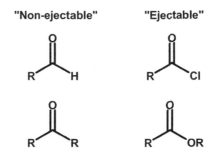

In terms of synthesis, we will then observe the following:

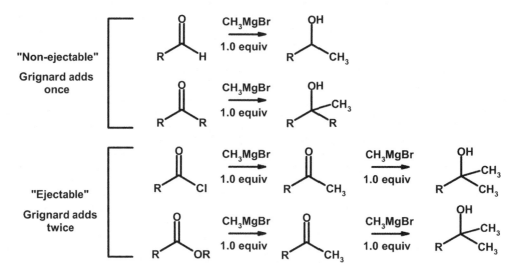

As shown above with the ketone and the aldehyde (which have non-ejectable substituents), the Grignard reagent can only add once to the carbonyl, giving an alcohol as the product in both instances. However, in the case of the acid halide and ester (which have ejectable groups attached) the first equivalent of Grignard kicks off the ejectable group to give a ketone. This ketone can then be reacted with another equivalent of Grignard reagent to give the final product, which is a tertiary alcohol. In the example above, we have added two equivalents of Grignard to the starting material in two different steps. However, if the alcohol is your desired end-product, you can do this all in one step by adding two or more equivalents of Grignard reagent.

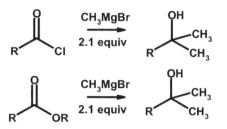

Take Home Message: Nucleophiles will substitute twice at the carbonyl if the starting material is an acid halide or an ester.

#43- Hint: "hv" means a radical is brewing.

This one is pretty simple. "hv" is Greek for "light and heat" and signifies that your reaction conditions contain light and heat added to the mixture. In nearly every instance in sophomore organic chemistry, this is the initiation of a radical reaction.

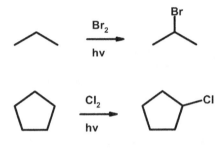

It is important to bear in mind that hv can significantly alter the reaction that occurs:

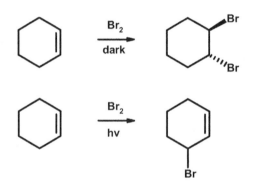

In the above example, the first reaction was the reaction of an alkene with molecular bromine to form the vicinal dihalide. In the second reaction, using the same reagents but different conditions, we undergo a radical reaction to provide the allylic bromide product. Remember that the only difference in these reactions is that the top reaction was done in the dark and the bottom one was done with the addition of light and heat.

However, you can also have a radical reaction without "hv" also, as peroxides are capable of initiating a radical reaction.

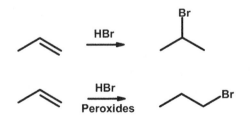

The first reaction is an ionic reaction where the product is based on the more stable carbocation intermediate formed. The second reaction is performed with peroxides and goes through a radical mechanism, where the more stable radical intermediate gives the major product.

Radical initiation is not limited to "hv" or peroxides though. Remember that heat or chemical initiators (such as AIBN) can all be used to initiate a radical reaction:

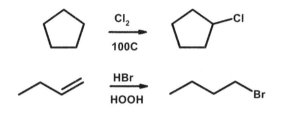

Take Home Message: If you see "hv", think radical mechanism

#44- T-butyl groups on cyclohexane means no ring flips....kinda.

We all know that cyclohexane undergoes ring flips between chair conformations. That is the easy part. It gets a little harder when talking about substituted cyclohexanes. The amount of ring flipping depends on what your substituent is.

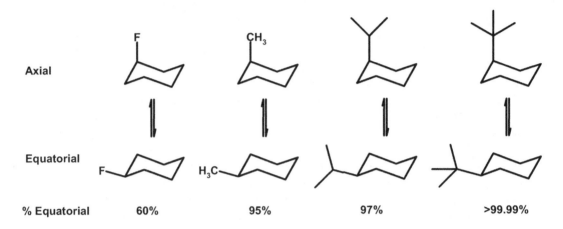

Axial				
Equatorial				
% Equatorial	**60%**	**95%**	**97%**	**>99.99%**

While the 1,3 diaxial interactions make the equatorial form more favorable, the amount of equatorial form in solution will depend on what the substituent is. As you can see above, at any one instant, 99.99% of t-butyl cyclohexane exists in the equatorial form. Does this mean that t-butyl cyclohexane is "locked" in the equatorial form? Yes and no.

Many professors will tell you that t-butyl cyclohexane is locked in the equatorial form and is never observed in the axial form. If this is what your professor wants you to know, then for your purposes, there is no axial t-butyl cyclohexane. However, in reality, axial t-butyl cyclohexane does exist in solution in minute quantities. We also know that E2 elimination cannot proceed unless the leaving group is anti-periplanar to the β-hydrogen. However, trans-t-butyl bromocyclohexane (where the leaving group is "locked" in a position that is not anti-periplanar) will undergo E2 elimination, albeit at a very slow rate.

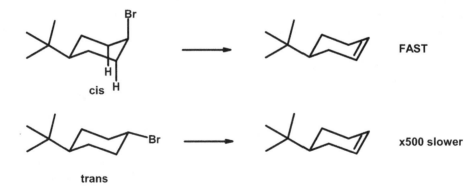

The fact that the reaction will proceed at all means that ring flipping it must occurring, even though it is "locked" with the t-butyl group. Therefore, t-butyl cyclohexane WILL ring flip, just at a very, very slow rate. This can also be seen in NMR studies.

However, an even better example of this is trans-decalin. Because of the geometry of the two fused rings, trans decalin will *never* ring flip. Therefore, you must be very careful if you see trans decalin in your exam. Usually it is only there to illustrate an E2 problem with no ring flip.

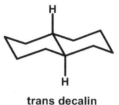

trans decalin

The problem below should illustrate this concept well. How many E2 products are available from the reaction below?

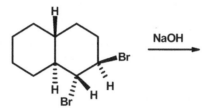

At first glance, one might think there are three possible E2 products, but once the structure is converted into its chair form, it becomes clear that there are no anti-periplanar hydrogens and therefore no possible E2 products.

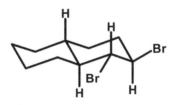

<u>**Take Home Message**</u>: **If your professor says t-butyl cyclohexane NEVER ring flips, then it NEVER ring flips. Otherwise, realize that it will ring flip at a very, very slow rate. Trans decalin NEVER ring flips.**

Bonus: How do draw a good cyclohexane ring.

Drawing cyclohexane properly is a huge challenge to organic chemistry students. Not only is it oddly shaped, but all of the bonds and substituents have to be in the right position to properly represent the molecule. This is going to be our brief lesson on how to draw it the right way. For your reference, bonds will be designated with letters.

Ring step 1: Draw two lines, at 45° angles to your paper. Lines a and b should be parallel to each other, and a decent distance apart.

Ring step 2: Lines c and d will connect lines a and b. Line c should be angled downwards, and should go about one-third of the way between lines a and b. Line d will be angled upward and will connect to line b.

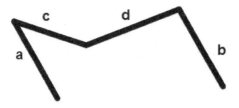

Ring step 3: Lines e and f are drawn similarly to c and d. Line e should be angled upward and go about 2/3 of the way between lines a and b. Line e should also be exactly parallel to line d. Line f will connect to line b and should be parallel to line c.

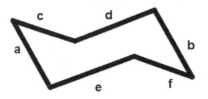

In the end, line a is parallel to line b, line c is parallel to line f, and line e is parallel to line d. This is important in order to get a good-looking cyclohexane.

Axial substituents: The easiest way to draw substituents is to start with the axial ones. Pick any one of the carbon atoms (represented at the line intersections) and draw a line going straight up from it. Move over one carbon, and draw the next line going straight down. Continue around your cyclohexene ring alternating up and down axial substituents.

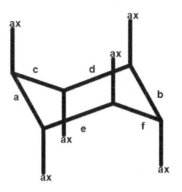

Equatorial substituents: This is another tricky part. Start with one of the carbons on line a or b, and draw an equatorial substituent. If you have done this correctly, they should be a little more than 90° from the axial substituents you just drew.

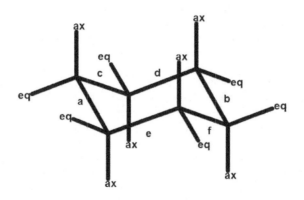

The perfect cyclohexane ring will have a tetrahedral carbon at each intersection. Everyone knows this is really difficult, but that is what you should be aiming for.

#45- Like Jerry Lewis asking for money, Lewis acids are always asking for electron density.

Lewis acids are like little burglars. They are always trying to swipe someone else's electron density because they are lacking it themselves. Organic chemists use this to their advantage, in that Lewis acids can be used to make electrophiles even more electrophilic. Because Lewis acids are electron-deficient, they will seek out electron-rich atoms. The electron-rich atom (the oxygen atom below) will then steal density from the closest atom to it (the carbon atom), which results in the carbon becoming more electrophilic. This is very much like "robbing Peter to Pay Paul" and will make the carbonyl carbon more prone to attack by a nucleophile.

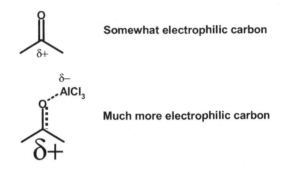

You have seen numerous examples of this in your organic chemistry class, as shown below:

Reaction	Example	Lewis Acid Complex

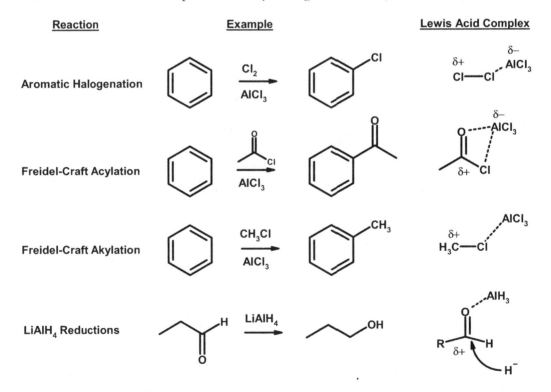

One way to think about this is in terms of bullies. Let's take the LiAlH₄ example. The Lewis acid is the biggest bully and steals lunch money from the oxygen. The oxygen is a bit of a bully himself, though. The oxygen then steals the same amount of lunch money from the carbon. Carbon is not a bully at all, but is actually a very sweet little boy. Carbon doesn't steal from anyone and is just left with no lunch money.

<u>Take Home Message</u>: Lewis acids steal electron density to make some atoms more electrophilic. If you see a Lewis acid in a mixture, there is probably an attack at the electrophile.

#46- H₂SO₄ and HNO₃: The good-cop/bad-cop of nitrations.

This one might be a simple one for most of you, but many of you will see this question on an exam somewhere, so it is worth discussing one more time. H_2SO_4 and HNO_3 combine together in the following way to form "NO_2^+", the nitronium ion, a highly electrophilic species capable of nitrating certain nucleophiles.

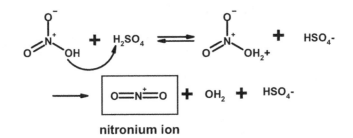

nitronium ion

Because sulfuric acid is a stronger acid than nitric (H_2SO_4 pK$_a$ of -4 vs. HNO_3 pK$_a$-2), the sulfuric will protonate the nitric acid. Once protonated, the nitric acid has a great leaving group attached to it (H_2O^+). The water is now capable of leaving, which generates the "NO_2^+", which will be the nitrating agent.

The most common example of this you will see is in electrophilic aromatic substitution (EAS) reactions:

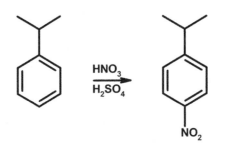

However, there are other times when this reagent can be used.

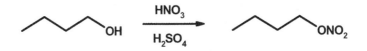

Remember that ultimately nitration is the process of a nucleophile attacking the nitronium ion and that many non-aromatic nucleophiles can also do this.

Take Home Message: HNO₃/H₂SO₄ is a nitration reaction.

#47- The Incredible Bulk: Bulky bases.

There are a number of bases that, when speaking in a steric sense, are very, very large. These "Basic Big Boys" are the baddest bullies on the block. Because of their size and bulk, they cannot react at most electrophilic sites on another molecule. This means that they can only act as a base, because they are too large to act as a nucleophile.

Below are the four most common bulky bases:

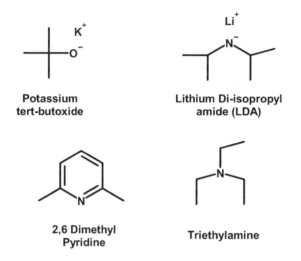

| Potassium tert-butoxide | Lithium Di-isopropyl amide (LDA) |

| 2,6 Dimethyl Pyridine | Triethylamine |

In substitution/elimination problems, elimination will always be favored over substitution when using these bases, as it is impossible for them to act as a nucleophile.

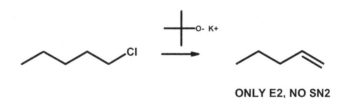

ONLY E2, NO SN2

Take Home Message: Bulky bases cannot act as a nucleophile, EVER!

#48 – Hoffman vs. Zaitsev: The elimination.

It is the match that we have been waiting for the ultimate, extreme elimination between Hoffman and Zaitsev. Here is how we break down the fight:

Hoffman	Zaitsev
Born: July 18, 1937	Born: 1841
Weight: Bulky	Weight: Slender
Rule: Proton abstraction occurs to give less substituted alkene	Rule: Proton abstraction to give more substituted alkene

There are generally two factors that will favor one fighter's product over the other. Factor one: Bulkiness of the base.

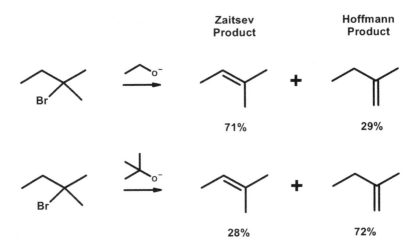

Tip #47 provided a summary of what is a bulky base and what is not. As you can see, the product obtained is greatly affected by which base you use. The general rule is that bulky bases will give a Hoffman product because the base is too large and cannot easily reach any proton that is not near the edge.

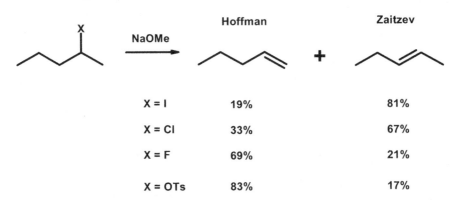

Leaving group is the second determining factor in these eliminations. The above reactions are all conducted using sodium methoxide, a very _non-hindered_ base. While it's no surprise that iodides and chlorides favor Zaitzev elimination, it should be somewhat surprising that tosylates favor Hoffman elimination and more surprising that fluorides are leaving groups at all. In fact, tosylates are large and bulky and block the place where the base would react to give the Zaitzev product. Fluorides are not large, but are such poor leaving groups that they do not allow the standard E2 elimination to proceed and are actually thought to proceed through another mechanism called E1cb. Don't worry, most professors will not require you to know the E1cb mechanism, which is thought to proceed through a carbanion intermediate.

Take Home Message: Bulky bases, "crazy" leaving groups, <u>or</u> unimolecular eliminations mostly lead to Hoffman products. Regular bases, normal leaving groups, <u>and</u> most bi-molecular eliminations (i.e. E2) give Zaitzev products.

#49- Dude, where's my carbocation?

There are few hard and fast rules in organic chemistry, but here is one of them: If a carbocation can rearrange, it will. Carbocation stability is: $3^0 > 2^0 > 1^0 >>$methyl. Therefore, if a 2^0 cation can rearrange to a more stable 3^0 cation, it will.

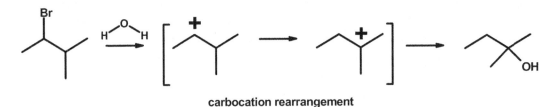

carbocation rearrangement

A better question to address here though is how do rearrangements occur. There are three mechanisms by which a carbocation can rearrange from less stable to more stable.

1. **1,2 Hydride Shift**: This is where H⁻ shifts from an adjacent carbon to the carbocation. This leaves a "+" charge on the carbon which lost the H⁻ and completes the hydride shift.

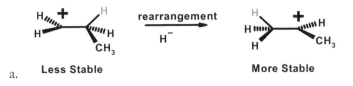

    a.   **Less Stable**   **More Stable**

2. **1,2 Methide shift**: The process is the same as with hydride shift, except in this case it is CH₃⁻ shifting instead of H. This is not as preferred as a hydride shift, but can still occur.

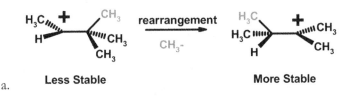

    a.   **Less Stable**   **More Stable**

3. **1,2 Phenyl shift**: Rare, but still possible are phenyl shifts, where an adjacent phenyl group shift to form a more stable carbocation. As with all carbocation formation, but especially with phenyl shifts, highly ionizing solvents are necessary for carbocation formation.

#50- Free radical halogenation: The molecular handle.

Let's face it, there is not much you can do with alkanes. They are decent nonpolar solvents, they power our SUVs/hybrids, and fuel our lighters. Alkanes are generally very unreactive because they are only composed of C-C and C-H single bonds, neither of which are very reactive. The issue for you as a sophomore organic chemist is that many of your professors will assign problems where you can only use a straight chain alkane as the starting material. Since the C-H bond is so unreactive, where do you start?

The best way to start one of these problems is to put a functional group on your alkane that is reactive. Since our choices are limited, we recommend starting with free radical halogenation. We believe that the best way to implement the free radical halogenation reaction is to use a starting material that is symmetric. This will provide a single product and avoid multiple products, which many professors will dock points over. Hence, the starting materials which will work the best are methane, ethane and all cycloalkanes.

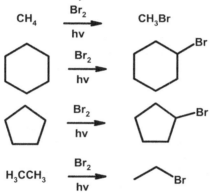

Bromination also provides you an addition advantage, as bromine is less reactive than chlorine. This means it is more selective for the more substituted carbon, which can support the radical intermediate more effectively. This allows the below reactions to take place in high yield.

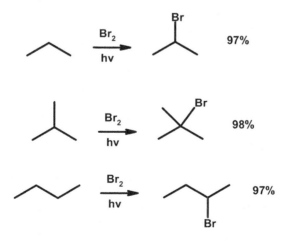

Halogenation provides a gateway into many, many other transformation, including substitution, elimination, oxidation, and organometallic formation. If one of these (or another)

transformation of the halogen is planned in subsequent steps and you don't need a specific halogen in the product, we recommend using the bromination reaction, as it is more selective and can be used with greater flexibility than chlorination.

Take Home Message: Know the alkanes which will give you one product via free radical halogenation, especially brominations.

#51- My Three Reductions: Reduction methods for alkynes

As we learned above, alkynes are a great way to form carbon-carbon bonds. Once you have performed your alkyne-carbon coupling reaction, what do you do with your triple bond if it is not in your final product? You have three alternatives:

- Partial reduction to the trans alkene: Using sodium and ammonia, you can reduce an alkyne to the trans alkene. The reaction does not work on alkenes, only alkynes, which is why the reaction does not continue the reduction to the alkane. The mechanism of the reaction is a single-electron transfer.

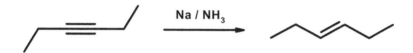

- Partial reduction to the cis alkene: Using "Lindlar's catalyst", you can selectively reduce the alkyne to the cis alkene. Lindlar's catalyst is $Pb-CaCO_3-PbO$ with H_2, however most professors will let you get away with just writing "Lindlar's catalyst" or "Lindlar Hydrogenation"

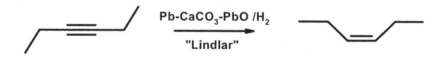

- Full reduction to the alkane: This is the nuclear option. This is also the hydrogenation that most of you are used to seeing. It is your standard hydrogenation using $H_2/Pd-C$. This will transform the alkyne to the fully-reduced alkane. This is a great way for making carbon-carbon bonds when you do not want to have any trace of the alkyne in your final product.

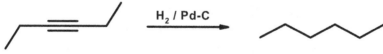

Take Home Message: Alkynes are a great way to form carbon-carbon bonds. They can then be reduced three different ways if you do not need a triple bond later.

#52- Is a halogen squatting on your molecule? Removing the unwanted halogen.

Just like in a late-night infomercial, where you might want to remove unwanted hair in your nose or on your upper lip, sometime you might want to remove that unwanted halogen from your molecule.

In the last section, we discussed adding a halogen to your alkane to give a "molecular handle" which is a place for your reactions to get started. But what happens if you get to a part of your synthetic sequence and have an extra halogen? How would you get rid of it? Can you just make them disappear? The answer is yes, through the magic of the Grignard reaction.

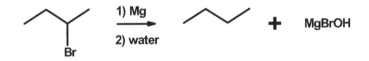

The reaction proceeds by formation of the Grignard reagent, which then attacks water and abstracts a proton, leaving MgBrOH salt as a byproduct. The reaction is highly exothermic, so care should be taken when conducting this reaction in the laboratory.

While this is a very simple example of halogen removal, this method can be applied to a number of situations where it is just time to have that halogen go away.

As shown above, this reaction is used to intentionally remove a halogen. For the same reason, when you are in your lab section forming a Grignard reagent, it is important to keep the reaction mixture free of moisture. Any water in your reaction will destroy your Grignard before it can be used.

Take Home Message: Remove the unwanted halogen with magnesium and water.

#53- You don't want a "D" on your transcript, but you might want one on your molecule.

Whenever scared organic chemistry students hear about a "D", they immediately run for the hills or curl up in the fetal position and hope it will all work itself out on its own.   But in organic chemistry, a "D" can be a wonderful thing. We look at "D's" as helper atoms sometimes. "D" is obviously deuterium, which is a hydrogen atom with one more neutron in its nucleus. Think of deuterium as a fat hydrogen. And because it is fat with the extra neutron, it can be detected in NMR experiments and can be used to determine the mechanism of some reactions. While this may not be fascinating to you, what should catch your interest is how to form these.

The reaction itself is easy and is very similar to Tip #52. All we are going to do is place a halogen where we want our deuterium to go, then react it with Mg/D_2O. D_2O is deuterium oxide, aka "heavy water", and will result in a net replacement of the halogen with deuterium.

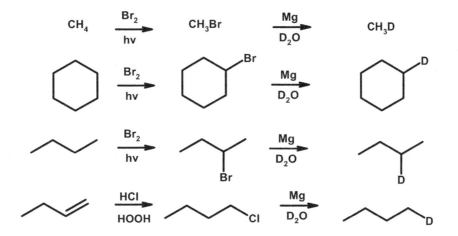

Warning: When conducting these reactions, be careful of chiral halides. Chiral halides do not necessarily form chiral Grignard reagents and therefore can lead to racemized products. Generally, when the reactions are performed with powdered Mg, the resulting product after treatment with D_2O is mostly racemic. While research has shown that using other forms of Mg (i.e. Rieke Magnesium) can give chiral deuterated products, it is not a given that your professor will recognize this and it is better to stay away from this reaction if the product is a chiral deuteride.

Take Home Message: If you are asked to create a molecule with a "D" somewhere in it, odds are it came from the halogen.

#54- Diels-Alder Part 2: Regiochemistry and stereochemistry

In Tip #30, we had a brief discussion about the Diels-Alder reaction. The take home message from this was that if you have a cyclohexane or cyclohexene ring, it probably came from a Diels-Alder reaction. Tip #30 had the basics of the reaction; now we will expand upon this. First, if you have a di-substituted dieneophile, it is important to note whether it is cis or trans, as the products from both will differ. The rule of thumb is that if you have a cis dieneophile, you will have a cis product at those substituents. Conversely, a trans dieneophile will give you a trans product, as shown below.

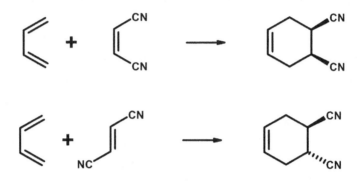

Further, if the diene is not symmetric, there is the chance for two different regiochemical products, as shown below:

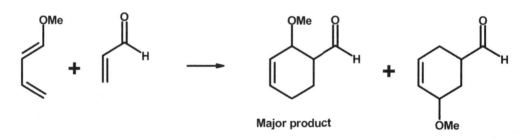

Major product

Why is the major product the "ortho" isomer shown above? [Note: We are calling this the "ortho" isomer because the substituents are next to each other. To be proper, ortho should only be used when speaking about aromatic rings, but we are going to cheat a little on that nomenclature here to make it easier to understand.] The answer lies in resonance structures. If we were to draw a resonance structure we would see that the diene has an electron donating group at the top, which places a partial negative charge on the furthest carbon. We can also draw a resonance structure for the dieneophile, which places a partial positive charge on the furthest carbon. Because we are certain that opposites attract, if we were to match the positive and negative portions of the reactants, we can easily see why the major product is the "ortho" product.

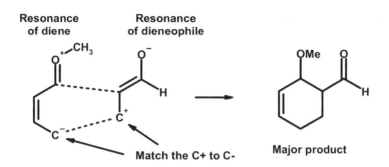

Major product

By the same token, we can look at another Diels-Alder reaction where the electron donating group on the interior part of the diene. In this reaction, the major product is the "para" product.

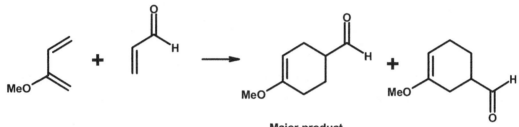

Major product

After drawing our resonance structures again, we can see that the same logic holds true for why we obtain this regiochemistry in the product.

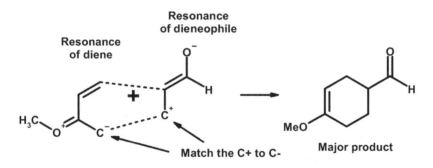

Major product

Finally, we have the endo-exo issue. Like the Pepsi Challenge, both the endo and the exo transition states look good, but only one can be the winner. In almost every Diels-Alder reaction you will see, the endo product is the major product of the reaction.

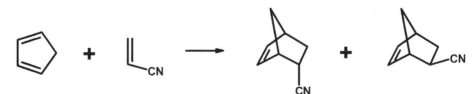

Endo isomer
Major product

Exo isomer
Minor product

#55- Diels-Alder Part 3: What makes a good Diels-Alder reaction?

While it's easier to get the Diels-Alder reaction to work than it is a Teamster, there are conditions that are necessary or can be optimized in order to achieve the best results. The first of these conditions is that the diene _must_ be able to adopt an "s-cis" conformation easily. The s-cis conformation is where the alkenes are on the same side of the central _single_ bond in the diene. If the diene is in the s-trans position, the Diels-Alder reaction will not work.

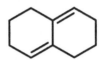

**a rigid diene that
cannot adopt s-cis
conformation = NO Diels Alder**

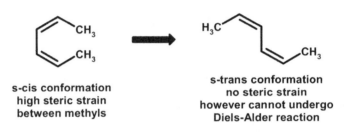

**s-cis conformation
high steric strain
between methyls**

**s-trans conformation
no steric strain
however cannot undergo
Diels-Alder reaction**

Second is a rule that is more common. The diene should have electron donating groups on it and the dieneophile should have electron withdrawing groups on it. For a list of common donating and withdrawing groups, see tip #32.

To expand upon this, remember that dienes and dieneophiles can come in all shapes and sizes. Do not forget that you can have a Diels-Alder reaction with something other than the bland, plain diene/dieneophiles that you may be used to. Below is a chart of some of the other ones that can work:

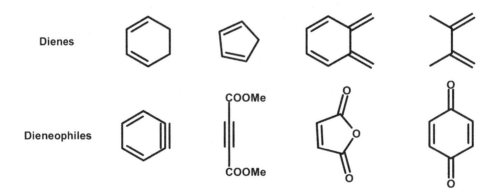

Last, the diene must not be too sterically hindered. Placing bulky alkyl groups on the ends of the diene are a sure way to stop a Diels-Alder reaction from working.

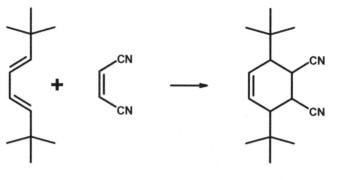

Poor Yield

#56- It's an urban legend that S$_N$1 reactions give _completely_ racemic products.

We have all heard the stories in our day:

"After stumbling in drunk the last night, the startled co-ed awoke the next morning to find a dead roommate and writing on the mirror that said 'You are lucky you did not turn on the lights'."

"….the police called back to say 'We have traced the call and it is coming from inside the house. Get out now!' But it was too late for the portly babysitter and her struggling artist hippie boyfriend, the killer had already found them."

And while we know better deep down inside, you still wonder sometimes about these urban legends. Well, today we will de-bunk one of these myths: S$_N$1 reactions give completely racemic products.

S$_N$1 reactions start harmlessly enough. We all know that the leaving group breaks off of the staring material, giving a carbocation intermediate which is then attacked by the nucleophile to give the final product.

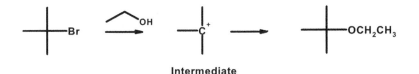

Intermediate

Upon further examination of the mechanism, we see that the carbocation intermediate has a trigonal planar geometry. This allows the nucleophile to attack from either face and give what we would believe is a completely racemic product. This is not the entire story.

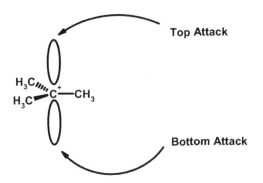

Top Attack

Bottom Attack

We know that the formation of the carbocation is a reversible step in S$_N$1 reactions, which means that the leaving group can come on and off of the starting material. But what happens when the leaving group is off, but not very far away? Could the leaving group partially block one face of the starting material? The answer to that question is yes.   To demonstrate this, we will use a chiral starting material to form our cation.

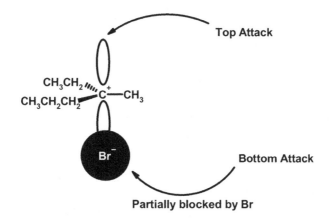

The leaving group usually remains close to the starting material, only finally departing once the reaction is completed. This partially/temporarily blocks one of the faces of the carbocation, the bottom face in this example. This leads to a slight preference for the product with inverted stereochemistry over the product with retention of stereochemistry. Chemists refer to this phenomenon as solvent-separated ion pairs. While it is very important to recognize the racemization of the starting material _DOES_ occur in S_N1 reactions, in reality it is not full and complete racemization. Typically, there is a 5-10% preference for the inverted product, meaning that the starting material is approximately 85-90% racemized.   Below are some examples:

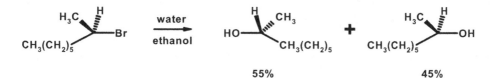

Take Home Message: S_N1 reactions leave you with a MOSTLY racemized product.

#57- "What do you 'Amine' you're getting rid of me?"--How to eliminate an amine.

Throughout organic chemistry you have been told that some atoms/groups are good leaving groups (See tip #18) and some are not. Normally, you would not consider an amine to be a good leaving group. However, this is not most circumstances. Enter the Cope Elimination. In this reaction, the amine is oxidized using hydrogen peroxide, then eliminated to the alkene with heat.

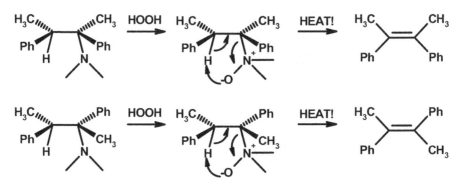

It is important to note that the stereochemistry of the alkene is determined by the stereochemistry at the chiral centers of the starting material. Further, there also must be a free *vicinal* hydrogen which can participate in the elimination.

Another example of where amines can be eliminated is seen in the Hoffman elimination, shown below:

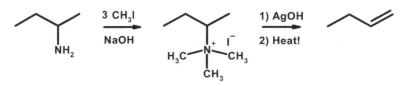

The reaction begins with exhaustive methylation to give the quaternary amine. This can then be eliminated using AgOH (used to replace I- with OH-), and heat to give the alkene.

This displays two very important features of both of these reactions. First, because amines are so nucleophilic, methylating four times to get the quaternary amine is an easy reaction to accomplish. Second, because the Hoffman elimination operates under Hoffman's rule (see Tip #48), which states that sterics have the greatest influence on the reaction, hydrogen loss will come from the least substituted carbon. This results in the less substituted alkene as the major product.

Take Home Message: Amines can be turned into leaving groups, and eliminated using the Hoffman or Cope elimination

#58- "Why Didn't You Say Were Sendin' the Wolf?"—The Wolff-Kishner reduction

The most likely place for this reaction is under the "making an unwanted functional group disappear" section. The Wolff-Kishner Reduction, just like the Clemmensen Reduction, removes carbonyl groups and converts them to the corresponding CH_2. The similarity between the two of these reactions is that both will only work on a ketone and aldehyde, not carboxylic acids. The difference is that Wolff-Kishner is performed under basic conditions, while the Clemmensen is performed under acidic conditions. Some examples of both are shown below:

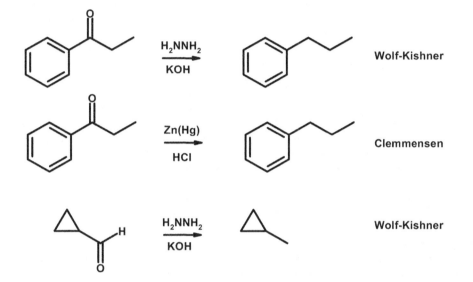

The Wolff-Kishner Reduction (the first and third reactions above) is accomplished using hydrazine, H_2NNH_2, and a strong aqueous base, which is usually KOH. [**WARNING**: Hydrazine is the chief component of space shuttle fuel and is *highly* toxic.] The Clemmensen Reduction is conducted using a zinc-mercury amalgam, and a strong acid, usually HCl. In most cases, the two reactions can be substituted for each other. However, be careful not to subject a base-sensitive molecule to Wolff-Kishner or an acid-sensitive starting material to the Clemmensen, or side reactions will occur.

While the mechanism of the Clemmensen Reduction is outside of the scope of an undergraduate organic chemistry class, the mechanism of the Wolff-Kishner Reduction is not. If we were to examine the components of the reaction, we would quickly recognize that carbonyls are electrophilic. Further, hydrazine is a very strong nucleophile. This means that the hydrazine will attack the carbonyl and provide a hydrazone as the intermediate of the reaction. The hydrazone is then degraded using KOH to give the final product. To be more specific, the base abstracts a proton from the free amine of the hydrazone, which is then passed to the carbon to be reduced. The driving force behind this degradation is the loss of N_2.

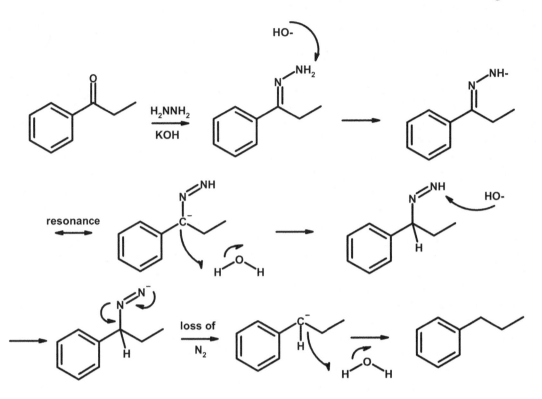

While this mechanism may look complex, it can be broken down into two separate portions: attack of hydrazine on the carbonyl and degradation in order to release N_2. Hydrazine is a very strong nucleophile, and thus will quickly attack the carbonyl. We know from previous examples that carbonyls and amines reacted together form imines. Next, our base induces several proton transfers which result in the ultimate loss of nitrogen, which gives us our final product.

Take Home Message: Wolff-Kishner and Clemmensen Reductions are both reactions that will convert a carbonyl into a CH_2.

#59- It takes a Baeyer-Villiger to raise a child

Esters can come from a number of reactions. Some examples of ways to form them from alcohols are shown below:

From acid chlorides

Fischer esterification

From carboxylic anhydrides

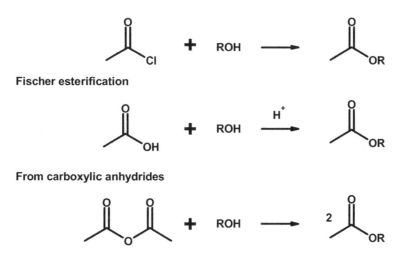

One of the very easy ways to form an ester is by using the Baeyer-Villiger reaction. This reaction takes a ketone and converts it directly to the ester through reaction with a peroxyacid.

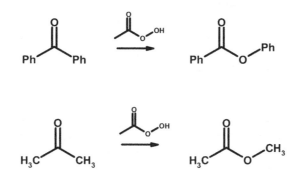

The examples shown above are symmetric ketones that would provide the same product regardless of which side the oxygen was added to. When the ketone is not symmetric, the oxygen will add to the more substituent side of the ketone. This means that the order of substitution will be tertiary before secondary before primary.

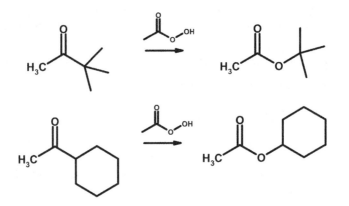

However, the main use of the Baeyer-Villiger reaction is in the formation of lactones (aka cyclic esters). These are not easily formed using the methods above (acid chloride, anhydride or Fischer methods) and are well-suited for the Baeyer-Villiger reaction. As above, the oxygen atom will still add to the more substituted side of the cyclic ketone.

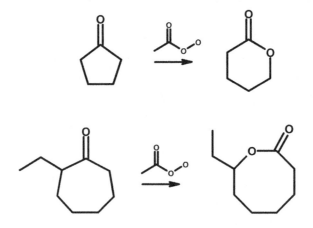

Baeyer-Villiger is generally regarded as the best way to create a lactone on an undergraduate exam.

Take Home Message: Cyclic esters (aka lactones) most likely came from the Baeyer-Villiger reaction.

#60- Symmetric diols came from the Pinacol reaction

Some of your professors will try to sneak this one in on you during an exam: Your professor will ask you to synthesize the symmetric diol shown below and you will either not know how or you will come up with a convoluted, difficult or wrong answer.

Synthesize this molecule using any organic molecule of TWO carbons or less

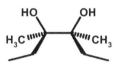

While not obvious, there is a very simple way to create a symmetric diol using the Pinacol reaction. Using a number of reducing agents, ketones and aldehydes can be coupled with themselves to form a symmetric diol, as shown below:

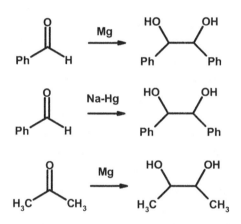

The reaction is also called the Pinacol Coupling Reaction and can be used on most aldehydes and ketones, but not on acid halides or carboxylic acids. Now to address the original synthetic problem above:

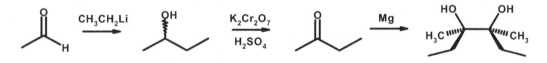

The first step is reaction of the reaction of ethyl lithium with ethanal to form 2-butanol. This is then oxidized to 2-butanone using Jones' oxidation. Now, under Pinacol Coupling conditions, 2-butanone is then reacted with itself using magnesium to form the final product.

Take Home Message: Symmetric 1,2 diols came from the Pinacol reaction.

#61- Convert a nitro group into the aldehyde or ketone.

Most of you are aware that nitro groups can be converted into amines via reduction. This reaction is most useful when performing electrophilic aromatic substitution reactions. However, nitro groups can also be converted into ketones and aldehydes under relatively simple conditions. While this reaction may not ever get used on a sophomore organic exam, it is good to know that it exists. The little-known organic reaction is the conversion of a nitro group directly into an aldehyde or ketone using alkaline $KMnO_4$. The reaction will convert a primary nitro group into the aldehyde and a secondary nitro group into a ketone.

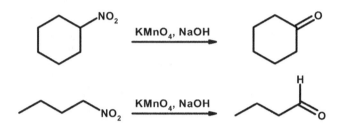

The reaction mechanism involves abstraction of the proton adjacent to the nitro group, followed by complexation of Mn to the oxygens of the nitro group, and rearrangement to the carbonyl compound. It is important to remember that $KMnO_4$ is a potent oxidizing agent, so this reaction should only be used on molecules where the $KMnO_4$ cannot unintentionally oxidize another functional group, such as –OH somewhere else on the molecule.

To go a step further, now if you wanted to completely remove the carbonyl, you could do that with the Wolff-Kishner reaction. (See tip #58). This now becomes a very effective way to completely remove a nitro group from your molecule if you so desire.

Take Home Message: When in a bind, you can convert an aliphatic nitro group into a carbonyl.

#62- The McMurry reaction: A "reverse" ozonolysis.

A handy reaction for the full oxidation of a double bond to a carbonyl is an ozonolysis reaction. Here, molecular ozone is used and goes through two intermediates, called primary ozonides and secondary ozonides. When final cleavage of the secondary ozonide occurs, it can proceed in two different paths, based on which type of workup is used. If a reductive workup is used (Zn, or CH_3SCH_3) then the resulting product will be an aldehyde. If an oxidative workup is performed (H_2O_2), the resulting product will be a carboxylic acid. Lesser known is that if a fully reductive workup is performed ($NaBH_4$ or $LiAlH_4$), the resulting product will be an alcohol.

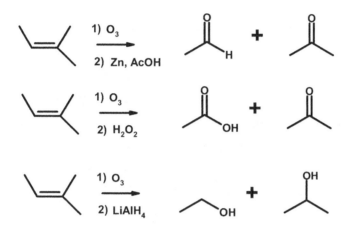

A quick trick to find all products of an ozonolysis reaction is to draw a dashed line through all of the double and triple bonds in your molecule and add an oxygen atom to the end of each bond you just cleaved. The type of functional group you create will be based on the type of workup. In our example below, we used a reductive workup with zinc, which will give aldehydes and ketones as your products.

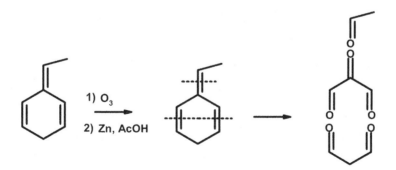

Much less known, however, is the McMurry reaction. At its essence, the McMurry reaction can be thought of as a "reverse ozonolysis" reaction. The reaction is performed using titanium trichloride, $TiCl_3$, (sometimes called "Tickle Chloride" or "Tickle Me Chloride") and will proceed in high yield under the right conditions. Those conditions are based on the starting material used, not the reaction conditions themselves. This means that only the right compounds will give good McMurry reactions.

There are two ways that the McMurry reaction can be successfully employed. First, is if the starting material is symmetric and is reacted with itself. In the McMurry reaction, all combinations of alkenes will be generated, based on the starting material. Therefore, if a symmetric starting material is used, only one product will result and a mixture of products can be avoided.

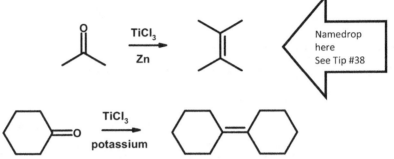

This reaction is not as effective when symmetric starting materials are not used, as cross-reactions are as plentiful as nerds at a Sci-Fi convention. See below for an example of a McMurry reaction with unsymmetrical starting materials.

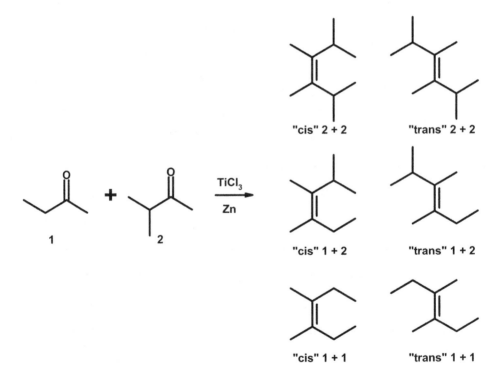

In the unsuccessful example above, there are six possible products. They are: starting material 1 reacting with itself, starting material 2 reacting with itself, and a cross reaction between 1 and 2, _WITH_ a cis and trans version for each. This should be avoided like it was a Sci-Fi convention. (Unsure why I am so down on science fiction today. Apologies to all Trekkies, Trekkers, and D&D kids.)

The other type of successful McMurry reaction is performed intramolecularly. The fact that the carbonyls are so close to each other in space allows for the reaction to proceed smoothly. Further, *inter*molecular coupling can be avoided if the reaction is performed under high dilution conditions. This can then be utilized as a very effective way of closing rings, especially large ones.

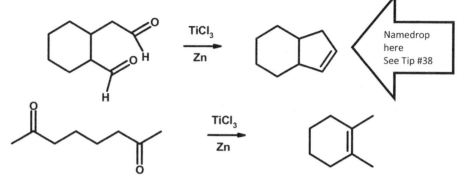

Help yourself out: As discussed in tip #38, it is a good idea to "drop some names", when using a reaction not taught by your professor. This will keep that snotty graduate student grader from marking you down for using perfectly valid chemistry.

Take Home Message: The McMurry reaction is a "reverse ozonolysis".

#63- Role reversal: Nitrile hydration and the Letts Nitrile synthesis

In the next installment of our series on converting two functional groups between each other, we now scrutinize nitriles. To do this, we will travel back in time and re-examine the forgotten art of electrophilic aromatic substitution (EAS). Nitriles can be easily converted into carboxylic acids by acid-catalyzed hydration, as shown below:

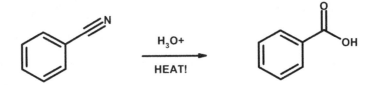

Further, if you were so inclined, you could go back to the nitrile from the carboxylic acid using a procedure called the Letts Nitrile Synthesis.

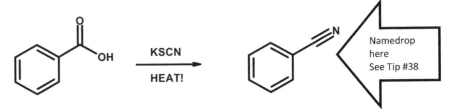

Using potassium thiocyanate, aromatic carboxylic acids can be converted back into the nitrile. This reaction is only valid for *aromatic* carboxylic acids, not for alkyl carboxylic acids.

The nice feature of this reaction, from an EAS perspective, is that the carboxylic acid and the nitrile are both meta-directors. Therefore, you can inter-convert between the two and without worrying about messing with your overall EAS strategy for installing different functional groups on your molecule.

Take Home Message: Nitriles can be converted into carboxylic acids via hydration and back to the nitrile using the Letts Nitrile Synthesis.

#64- Role reversal: Dihydroxylation and Corey-Winter Olefination

The more common reaction of this tip is the dihydroxylation, where a double bond can be reacted to form a diol, as shown below.

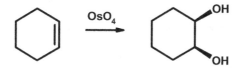

In this reaction, OsO_4 is used as the dihydroxylating agent, and gives the cis diol as the product. This reaction is very useful and is one of the major ways to form a cis 1,2 diol. In contrast, the trans 1,2 diol can be formed by epoxidizing a double bond, then opening it with hydroxide ion.

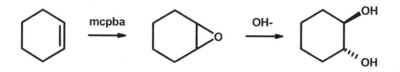

Yet there is a much lesser known way to go from a 1,2 diol back to the double bond. The reaction is called the Corey-Winter Olefination and is a simple method for obtaining double bonds.

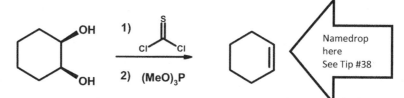

The mechanism of the reaction is a little more complex. It involves reaction of the diol with thiophosgene, to form the cyclic thio-carbonate. This is then degraded to the double bond using trimethyl phosphite. Since the reaction proceeds through a cyclic intermediate, *it can only be performed successfully on cis diols*. Therefore, we could NOT perform this reaction on the trans diol shown above.

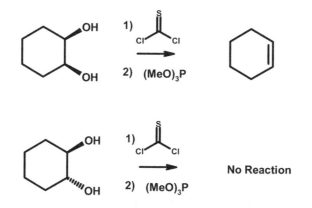

Remember that if the diol is a straight chain alkane, then it may be rotated to place the hydroxyls in a cis conformation, so that even though it appears to be a trans diol, the reaction can still be performed.

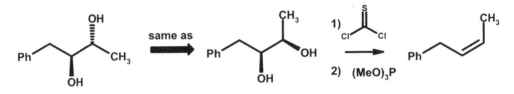

The configuration of the alkene obtained depends on the original configuration of the diol. Wherever the diol substituents are located in the original starting material is how the alkene will be formed.

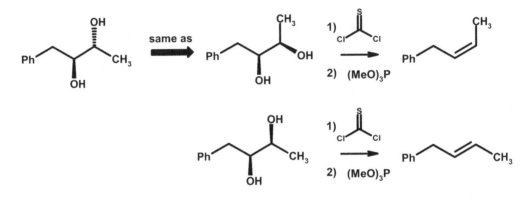

Take Home Message: OsO$_4$ can convert an olefin into the cis 1,2 diol. Thiophosgene/trimethyl phosphite will convert the cis diol back into the olefin.

#65- Role reversal: Halogens and carboxylic acids

As discussed in previous sections halogens are the most versatile functional group in organic chemistry. They are the Eddie Murphy of functional groups. They can be turned into any one of a number of zany characters, like double bonds, ethers, or carboxylic acids. The conversion to a carboxylic acid is what we will discuss here. The reaction is relatively simple, and consists of reaction of the halogen with magnesium metal followed by reaction with excess carbon dioxide. The magnesium is responsible for converting the halogen to a Grignard reagent, which is the reacted with carbon dioxide, an excellent electrophile.

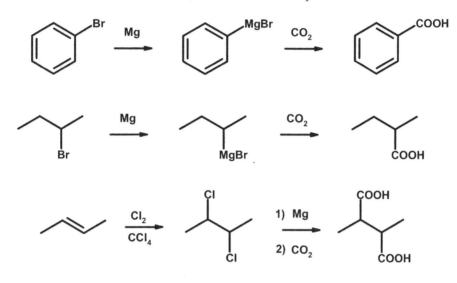

The mechanism of the Grignard attack is worth briefly discussing. The Grignard reagent is an excellent nucleophile. This will attack carbon dioxide, which is a very good electrophile. Once this reaction has occurred, the product is the magnesium bromide salt of the carboxylic acid, which must be converted to the free carboxylic acid. This can be accomplished using by acid workup.

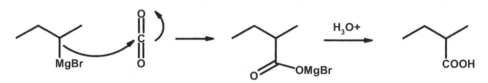

The reverse of this reaction, which can turn a carboxylic back into a halogen, is the Hunsdiecker reaction. In this reaction, the carboxylic acid is converted to the silver salt using AgOH, which is a strong base. This is then reacted with molecular bromine, to convert the compound to the bromide.

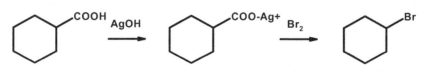

Above the reaction is shown in two separate steps, however on an exam you will be able to combine the two steps as shown below.

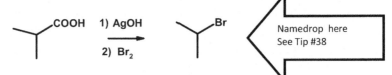

AgOH is a strong base, so be careful when performing this reaction in the presence of another functional group which it might react with. The obvious atom you don't want somewhere else on your molecule is a halogen, which could possibly eliminate to the double bond. You should also avoid any double bonds in your starting materials, as this will give poor yields in real life and may cause your professor to dock you points.

The reaction is best suited for alkyl carboxylic acids, but can also be used on aryl carboxylic acids. Further, the reaction is not solely limited to formation of the bromide, as the iodide and chloride have also been formed.

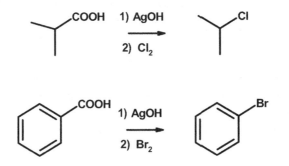

Finally, the reaction will only transform carboxylic acids; it will not affect any other type of carbonyl within your molecule, as shown below.

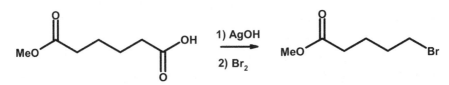

Take Home Message: Mg/CO$_2$ can convert a halogen to the carboxylic acid. AgOH/X$_2$ will convert it back to the halogen.

#66: Musical double bonds: Using catalytic I₂.

Once again, iodine just can't play like its other halogen siblings. You would expect this reaction to occur like di-bromination or di-chlorination, where the halogen adds across the double bond in an anti-fashion, but iodine doesn't play like that.

When catalytic iodine is subjected to an alkene, isomerization (not addition) occurs. The reaction starts out the same as bromination or chlorination, as the halogen reacts to form the iodonium ion (the first intermediate). However, instead of I⁻ opening up the three-membered ring to give the di-iodination, the ring opens on its own, placing a positive charge adjacent to the halogen. This is referred to as the epihalonium ion. I⁻ then acts as a base and removes the iodine and creates the new double bond. The reaction works because the iodonium intermediate does not stay a three-membered ring, but opens up and allows free rotation about the newly-formed carbocation, which allows the newly-formed double bond to assume the more thermodynamically-favored trans configuration.

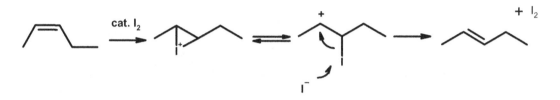

As you can see above, at the end of the reaction, we have regenerated molecular iodine, which is the reason we can use catalytic amounts of iodine.

This catalytic isomerization reaction is not confined to small molecules, but has also been used on large ones too. Below is the transformation from 11-cis-retinal to all-trans retinal. In the human retina, a photon of light may strike 11-cis-retinal and convert it into all-trans-retinal. However, we can induce this transformation in a flask in the dark by adding catalytic iodine, as shown below.

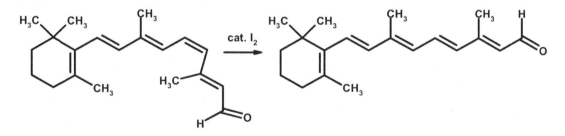

Take Home Message: Catalytic I₂ does not behave like other halogens in the presence of alkenes, as it will transform cis double bonds into trans double bonds.

#67- Just Like the Secret Service, Protecting Groups Are Chemical Bodyguards

Functional groups are the VIPs of the chemical world. They have all of the influence in a given reaction and wield all of the power. Consequently, there are a number of bad characters out there that want to assassinate the functional groups. These are reagents that are meant to react with one functional group, but could "kill" another functional group. Below is a classic example of this:

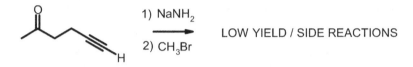

1) NaNH$_2$

2) CH$_3$Br

LOW YIELD / SIDE REACTIONS

The intended reaction is the deprotonation and alkylation of the alkyne above. However, this reaction will not occur in high yield. This is because once the nucleophile is created, it will react with the closest electrophile (the carbonyl) and not the intended one (the CH$_3$Br). Once again, our VIP has been assassinated. Next time, let's get a bodyguard.

Below is another reaction sequence that may show this concept a little more clearly. If we just reacted the starting material with LiAlH$_4$, we would reduce both the ketone and the carboxylic acid. However, we can selectively reduce the carboxylic acid if we wanted by protecting the ketone. The sequence is the same as above. We will protect the ketone as the 1,3-Dioxane. We can then selectively reduce the carboxylic acid with LiAlH$_4$, and then remove the protecting group with strong acid. This example should clearly show that the results are quite different when we use a protecting group.

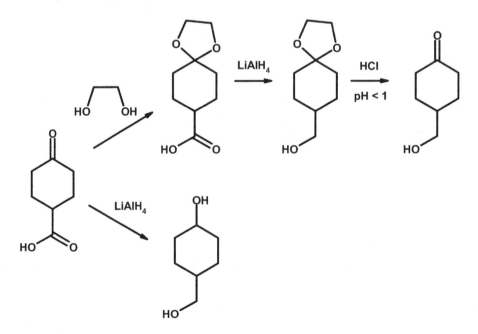

To speak more specifically about protecting groups, there are three generic steps to using a protecting group on your functional group. The first step is protection of the functional group (hire a bodyguard). The second step is to perform the desired reaction (bodyguard protects the VIP against attack). The third step is to deprotect the functional group to give the end product you want (pay the bodyguard and send him home). Armed with this knowledge, we can perform the first reaction the right way.

Protect

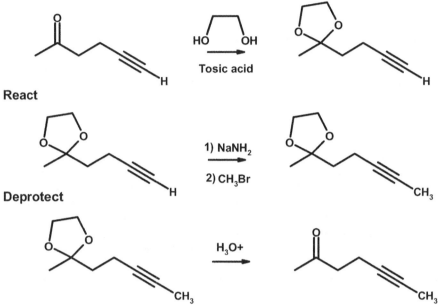

In the reaction sequence above, the carbonyl is protected using ethylene glycol and tosic acid as a catalyst. This installs the protecting group on the carbonyl so it cannot react. In this case, the carbonyl is protected as the acetal, which is a geminal di-ether. We know from previous sections that ethers are very unreactive, and therefore make a good protecting group. We are then free to perform our reaction, which is now selective for the alkyne. Finally, we will deprotect the carbonyl with acidic water. This returns the original carbonyl to us and provides our final product. Although this seems like an overly lengthy sequence, it is the best way to synthesize the final product.

So, what makes a good protecting group? The same things that make a good bodyguard.

1) It should be easy to install. Think of this in terms of the bodyguard analogy, meaning that our bodyguard has to be effortless to hire.
2) It should be resistant to a number of reagents that might normally affect our functional group. This means our bodyguard has to be tough and protect against a number of threats. It can't just protect against Al-Qaeda, but must also be able to protect against HAMAS, the mafia, and hippie protestors.

3) It should be easy to remove. After protecting against the threat, the bodyguard must be able to go home that night when the job is done, leaving the VIP like he was before the bodyguard was hired.

Functional Group to protect	Protected as	Protection reaction conditions	Deprotection reaction conditions
Alcohol [ROH]	Trialkyl Silyl ether	R_3SiCl, Pyridine	H_3O^+ (pH < 1) or HF
Alcohol [ROH]	Benzyl ether	$PhCH_2Cl$, Pyridine	H_2/Ni, or Na/NH_3
Carbonyl [RC(O)R]	1,3-Dioxane	Ethylene Glycol, tosic acid	H_3O^+ (pH < 1) or Zn/HCl
Amine [R_2NH]	Acetate	CH_3COCl, Pyridine	Na/NH_3

Entire books have been written on protecting groups, but we have summarized a few of the major ones above. We _STRONGLY_ encourage you to read up on these much more before you try to use this chemistry on your exam, as this is only meant to be an introduction to the topic.

On your exam, you will likely see a very simple problem using protecting groups. It will most likely only involve protection, followed by one reaction and immediate deprotection. Because of this, it is probably not a good idea to waste a lot of time learning protecting group chemistry. We suggest that you learn just a few simple protecting groups and spend more time learning other organic reactions.

Take Home Message: If you have two sites that a reaction could potentially occur at, protect one, react at the other, and deprotect the first.

#68- Don't hydrogenate that aryl ring! You'll kill us all!

In tip #30, we stated that the main method for obtaining a substituted cyclohexane was the Diels-Alder reaction. [Your professors love that reaction]. However, in an effort to expand your realm of thinking, we are now going to introduce to you a second way to make highly substituted cyclohexane rings. Suppose you are given this problem:

Synthesize the following molecule from organic compounds of six carbons or less

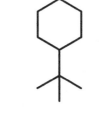

At first blush, you might think that the following synthesis would be valid:

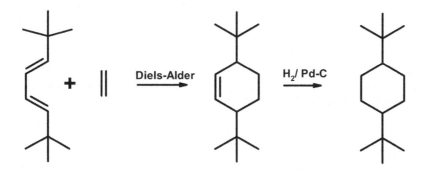

However, there are two problems with this synthesis: 1) Your starting material has more than 6 carbons, and 2) Diels-Alder reactions using dienes that have a t-butyl group can be difficult. Hence, we should have a second route in mind for situations like this.

An alternative method is to perform electrophilic aromatic substitution reactions on benzene, then reduce benzene to cyclohexane via a harsh hydrogenation. This method would be viable because the following reaction can be done in the lab:

When doing this reaction on an exam, be sure to note two things to get full credit. First, use Ni as your catalyst, not Pd-C. Second, be sure to somehow emphasize the fact that the reaction is performed using high heat and pressure. (We have chosen to highlight that here using all capital letter and placing an exclamation point at the end. This is also why we are worried that you will kill us all—high pressure hydrogen is like a small bomb waiting to go "boom".)

Using this method, our synthesis will look like this:

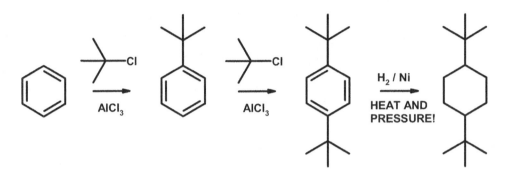

As stated above, the key is to do the EAS chemistry first, then reduce the aryl ring to the cyclohexane. The vast majority of professors will give you full credit for this reaction sequence provided that you emphasize the conditions under which the aryl ring is being reduced.

As you saw in your course, there are a number of different EAS reactions that can be performed on benzene. Some of these, like the t-butyl group above, will be inert to hydrogenation. However, others will be reduced to other functional groups under these reaction conditions. Some of these are shown below:

Function Group on the ring (Ar = C_6H_5)	During H_2/Ni reaction, converted to: (C_6H_{11} = cyclohexane ring)
Ar-CN	$C_6H_{11}-CH_2NH_2$
Ar-NO$_2$	$C_6H_{11}-NH_2$
Ar-C(O)CH$_3$	$C_6H_{11}-CH_2CH_3$
Ar-R	$C_6H_{11}-R$ (inert to H_2/Ni)
Ar-OR	$C_6H_{11}-OR$ (inert to H_2/Ni)
Ar-OH	$C_6H_{11}-OH$ (inert to H_2/Ni)
Ar-NH$_2$	$C_6H_{11}-NH_2$ (inert to H_2/Ni)
Ar-X	Will stop the reaction, no reduction of ring

Please note that this reaction cannot be conducted if there is a halogen on the aryl ring, as it will poison the catalyst and halt the reaction.

The table above shows the reaction that occurs when Ar rings are subjected to these harsh hydrogenation conditions and states that about half of your average EAS substituents will not be changed by the H_2/Ni reaction. In these reactions where the substituent is not affected, only the ring will be hydrogenated. Some of the substituents will be affected though. For example, nitrobenzene will be transformed to cyclohexylamine under these conditions. Here, not only is the ring being reduced, but the nitro group is being converted into an amine, as shown below:

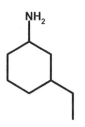

While hydrogenation may affect some of these groups in ways you did not think of, we can still use this to our advantage. Just keep in mind that during the last step of the reaction, you are going to transform some of these. For example:

Synthesize the following molecule from organic compounds of six carbons or less

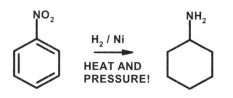

As shown below in the answer to this problem, we obtain the product by performing standard EAS chemistry, then reducing the compound in the final step. Notice that we needed NO_2 as a meta director in the second step, then reduced it to the amine in the final step.

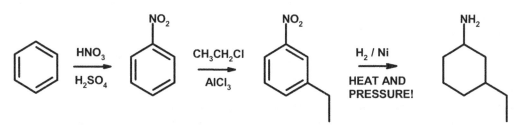

<u>Take Home Message</u>: Diels-Alder is the primary method for making cyclohexane ring on exams. However, if you get in a bind, consider doing EAS chemistry then reducing the aryl ring to a cyclohexane with H_2/Ni under high heat and pressure

69- It's as Easy as 1,2 Addition. Or Is It 1,4 Addition?

Everything goes back to resonance. And here we are again, back examining resonance structures. In this instance, we are looking at α,β-unsaturated ketones. If we study the resonance structures, we see that there are two different electrophilic sites: the bottom of the ketone and the back of the alkene.

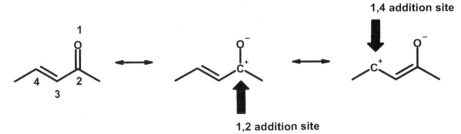

If we start counting from the oxygen atom, we can label all of the atoms from 1 to 4. Resonance structures of the unsaturated ketone above show that we do have two sites that a nucleophile could add to. The burning question now is which nucleophiles will add to which sites and why.

Adding to the carbonyl carbon is referred to as 1,2-addition. This is also called "direct addition". The 1,2-addition product is formed faster and is therefore the kinetic product. In general, strongly basic nucleophiles will undergo 1,2 addition. Some examples of these strongly basic nucleophiles are Grignards, organolithiums, and alkyne anions, as shown below:

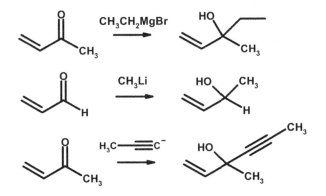

Addition to the alkene carbon is called 1,4-addition. This is also called "conjugate addition" or "Michael addition". The 1,4 product is formed slower, but is more stable and is therefore the thermodynamic product. In general, weakly basic nucleophiles will undergo 1,4 addition. Some examples of weakly basic nucleophiles are organocuprates, alkyl thiols, CN^-, most enolates, and amines, as shown below:

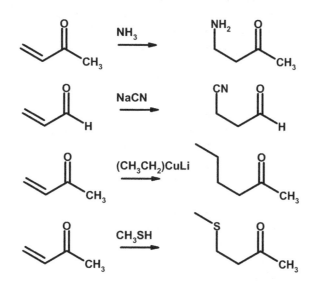

The trick here is to recognize that you can add the same group to two different spots on a molecule depending on which reagent you use. For example, if you wanted to add a methyl group via 1,2-addition, you would use methyl Grignard. Conversely, if you wanted to add the methyl group via 1,4-addition, you would use methyl cuprate.

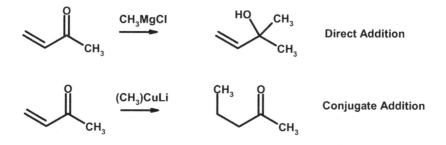

Take Home Message: There are two types of addition, 1,2 and 1,4 addition. Strongly basic nucleophiles participate in direct addition. Weakly basic nucleophiles participate in conjugate addition.

#70- Here's to your annulation, Mrs. Robinson: Musings on the Robinson Annulation.

As we just learned in the last tip (hopefully you haven't forgotten already) there are two types of addition to α,β-unsaturated ketones: direct and conjugate. In this tip, we are going to focus on a very special type of conjugate addition called the Robinson Annulation. This is a favorite of many professors and the most likely method you would see on an exam for synthesizing a compound that has two fused cycloalkanes, like the one shown below:

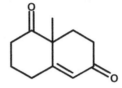

The Robinson Annulation has two main steps. The first step is a Michael (aka conjugate) addition between an enolate and an α,β-unsaturated ketone or aldehyde. Because enolates are weakly basic, they will add in a 1,4 fashion to the unsaturated ketone to give the product shown below. KOH is used as the base to form the enolate on the starting material.

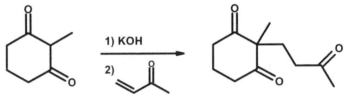

The second step is the cyclization step. Once again, we subject the starting material to base to form an enolate which will attack a carbonyl. There is a 1,2 addition to a carbonyl because there is no double bond for a 1,4 reaction. Therefore, the nucleophile defaults to a 1,2 reaction.

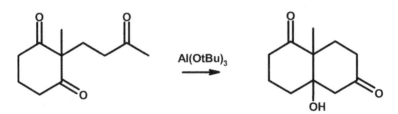

Upon closer examination, we can see that there are three different α-protons (next to the carbonyls) which could be deprotonated to give enolates which will attack a carbonyl. Moreover, there are two different carbonyls which could be attacked by the enolate. So why do we only get one product out?

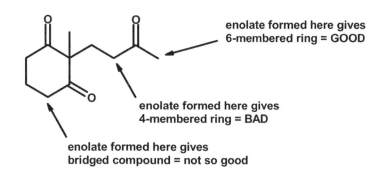

**enolate formed here gives
6-membered ring = GOOD**

**enolate formed here gives
4-membered ring = BAD**

**enolate formed here gives
bridged compound = not so good**

Once we dissect the three different places where an enolate could be formed, we see that there is really only one that would give a thermodynamically favorable product, the 6-membered fused ring.

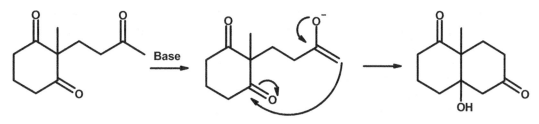

Base

Enolate Not Isolated

If we add a little heat to the final product, it will dehydrate to the unsaturated ketone. Now we can put the whole thing together, where it will look like this:

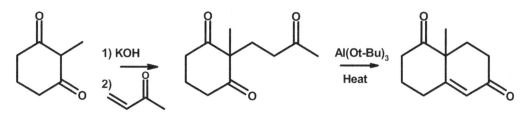

This is also not limited to fused 6-membered rings, but can also be used to form smaller fused ring systems.

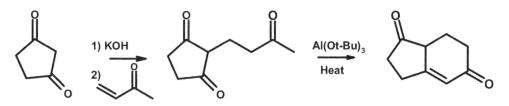

Finally, we can also use this method to make regular six-membered rings. This method is very useful for cyclohexane rings where the Diels-Alder reaction just won't work right.

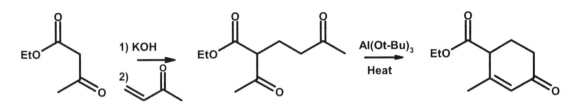

Please note that if we were to try to create this molecule via Diels-Alder, we would have to place a ketone and ester on our diene. This is highly unfavorable as the diene should have electron donating groups on it. Therefore, Diels-Alder is not the best method for making this compound.

Take Home Message: The Robinson Annulation can be used to create 6-membered rings, whether they are fused to another cycloalkane or alone as a cyclohexane.

#71- Kinetic vs. Thermodynamic Enolates: You Choose the Winner of This Fight.

An enolate ion can be formed when the α-proton of a carbonyl is subject to base. This removes the proton and creates an enolate anion, which is nucleophilic and can participate in S_N2 reactions. However, when dealing with asymmetric ketones, we see that there could be two different α-protons which could react with the base. Which one will react first?

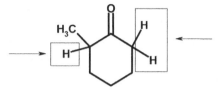

Two different sets of alpha protons

When examining the ketone above, we see that the protons to the right and left of the carbonyl are not equivalent and therefore may react at different rates. When discussing enolates, there are two different types of enolate ions that can form: the thermodynamic enolate and the kinetic enolate. The chart below summarizes the differences between the two.

	Kinetic Enolate	Thermodynamic Enolate
C=O substitution	Less Substituted Side	More Substituted Side
Temp formed at	Low temperatures (i.e. -78⁰C)	Higher Temperatures (> 0⁰C)
Base used	Strong, hindered base	Weaker, unhindered base
Reversible?	Irreversible reaction	Reversible reaction
Speed of formation	Forms quickly	Forms more slowly
Base counterion	Prefers lithium	Prefers sodium or potassium

The most important differences to take away from the table above is that the kinetic enolate forms quickly and needs a strong, hindered base at lower temperatures while the thermodynamic enolate is more stable and needs a weaker unhindered base at higher temperatures.

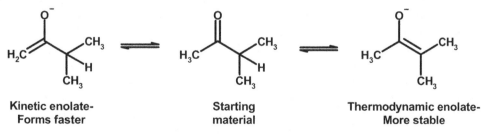

Kinetic enolate-
Forms faster

Starting
material

Thermodynamic enolate-
More stable

The figure below is a good example of kinetic control in enolate formation. Notice all of the conditions used favor the kinetic product (strong hindered base, low temps, lithium counterion).

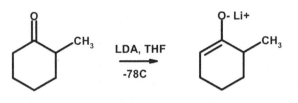

Below are several more examples of how changing the condition under which the enolate is formed will affect which product is obtained.

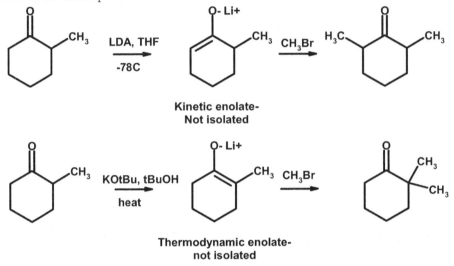

In the above example, the alkylation site is changed by preferentially forming one enolate over the other. In the example below, an entirely different ring system is obtained by using a different base and different temperatures. Below, instead of using CH₃Br to alkylate intermolecularly, we have the alkylhalide attached to our enolate making it an intramolecular alkylation.

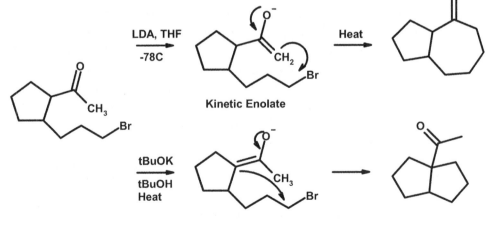

<u>Take Home Message</u>: Kinetic enolates are formed quickly and are formed with bulky bases at low temperatures. Thermodynamic enolates are more stable and are formed with weaker bases and higher temperatures.

#72- "Where's Aldol?" Making an Aldol Reaction Work for You.

Aldol reactions are not too hard to get a grasp on. Aldol products come from the enolates of aldehydes. In this reaction, we take advantage of two factors. First, because of the electron withdrawing effect of the carbonyl, the α-protons adjacent to the carbonyl are acidic. The resulting carbanion can then be used as a nucleophile. Second, the carbonyl itself is electrophilic, and can accept the electron density possessed by the nucleophilic carbanion. The interesting part of this reaction is that both the electrophile and nucleophile are on the same molecule. Therefore, if we take two of them together, they can react with each other to form a new product as shown below.

The mechanism of this reaction is relatively straight-forward. The base deprotonates the α-protons to give the carbanion, which is in resonance with the enolate ion. The enolate ion then attacks the carbonyl of the other aldehyde, which after protonation with acidic water, gives the final product.

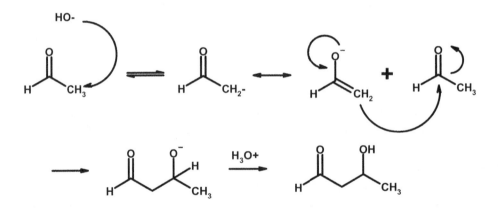

The above example was performed at 5°C and gave the saturated product, where the hydroxyl was still present. In the example below, the reaction was performed at 90°C and provided the α,β-unsaturated product. Dehydration occurs spontaneously at this temperature, and does not require a separate step.

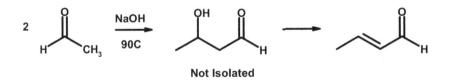

Not Isolated

In all of the examples shown thus far, we have been reacting one aldehyde with itself. But could we react two different carbonyl compounds to get a "mixed aldol" product? The answer is yes, but we need to be careful which carbonyl compounds we use. Below, we have a mixed aldol reaction which would be a disaster.

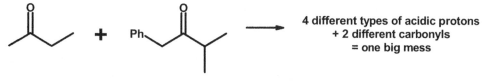

**4 different types of acidic protons
+ 2 different carbonyls
= one big mess**

There are some ways however to stop mixed aldol products from running wild on us:

1) <u>Limit the number of α-protons on your molecules.</u> In the below mixed aldol reaction, there is only one acidic proton that can react. Although there are two carbonyls, as we will see next, the aldehyde is more reactive.

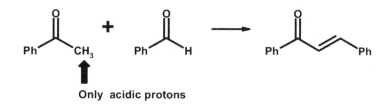

Only acidic protons

2) <u>Use a more reactive carbonyl.</u> Aldehydes are more reactive than ketones and formaldehyde is more reactive than other aldehydes. The nucleophile will preferentially attack the more reactive carbonyl.

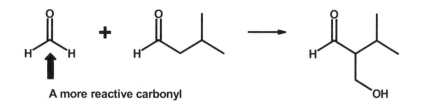

A more reactive carbonyl

<u>Take Home Message</u>: An aldol reaction is the reaction of an aldehyde with itself. Mixed aldols are possible, but to avoid multiple products you should limit the number of acidic protons and use a more reactive aldehyde.

#73- Acetoacetic Ester Synthesis: What you need to know in one page or less.

While this may look intimidating, the acetoacetic ester synthesis is the most effective method for created substituted methyl ketones. The basis of the method is to alkylate a β-diketone, taking advantage of the fact that the α-protons between the two carbonyls are extra acidic. A generic synthesis is shown below:

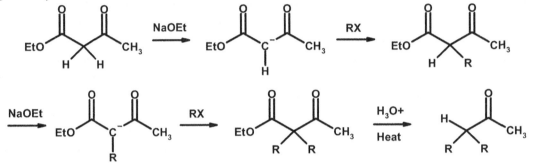

The essence of the reaction sequence is deprotonation/alkylation. In the above example, deprotonation is achieved with sodium ethoxide, while alkylation is accomplished with an alkyl halide on the resulting carbonion. The sequence is repeated twice to give di-alkylation. Finally, the ester is hydrolyzed and decarboxylated to give the substituted methyl ketone (aka substituted acetone).

Sometimes, the reaction will be shown with the acetoacetic ester and alkyl halide on one side of the arrow:

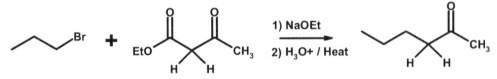

The most common version of this problem uses the acetoacetic ester above. However, any alkyl group can replace the methyl group so long as there is a β-ester on the other end that can be hydrolyzed and decarboxylated. In the below example, the methyl group is replaced by a cyclopentyl group.

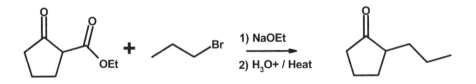

Take Home Message: If you see the acetoacetic ester in a problem, think deprotonation/alkylation and a decarboxylation when it is finished.

#74- β-keto Esters Came from a Claisen Condensation (NOT the Claisen Rearrangement)

The Claisen condensation is an aldol-type reaction, except it is between two esters. The other notable difference between the two is that while the aldol reaction has the nucleophile attack the carbonyl reducing it to an alcohol, the Claisen condensation "ejects" one of the esters (see tip #42 to talk about "ejectable" groups")

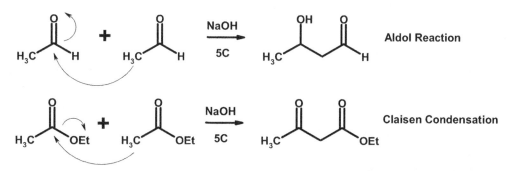

To restate this, in the aldol reaction, the enolate created attacks the other aldehyde carbonyl and converts it to the hydroxyl in the final product. In the Claisen Condensation however, the attacking enolate ejects the ester portion, forming a β-diketone product.

These problems can become more complex when dealing with longer chained esters, as shown below:

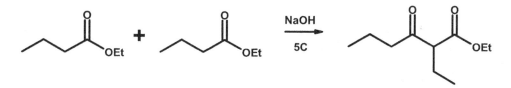

It is most important here to remember to count your carbons in your final product to make sure that you have not forgotten any of them. The key mechanistic step of this reaction is very similar to that of the aldol reaction. After deprotonation to form the enolate, the enolate ion attacks the other ester and kicks off the ester portion to form the final product.

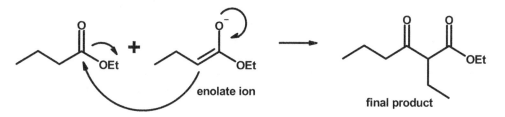

Just like two cousins that grew up together, the aldol and Claisen reactions have even more in common. Just like with the aldol reaction, you can also perform a mixed Claisen reaction. Again, the key is to limit the number of α-protons that can lead to enolates and have a more reactive carbonyl. In the reaction below we have both, as there is only one α-hydrogen that can be deprotonated and the carbonyl acceptor is a formic ester (aka aldehyde on one side, ester on the other).

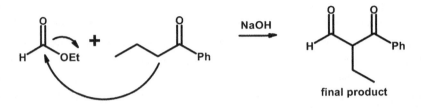

Further, both reactants do not have to be esters. The defining characteristic of the Claisen condensation is that an ester is ejected. The reaction works just as well when a ketone is the source of the enolate nucleophile.

Take Home Message: The Claisen condensation is similar to the aldol reaction, except there is an ester that gets ejected.

#75 – Convert ketones to epoxides fast

The general school of thought for most sophomore organic chemistry students is that epoxides came from double bonds as shown below:

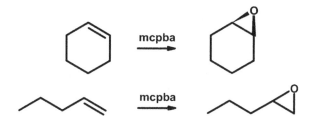

However, just like The Matrix did to Neo, we want to expand your thinking. We want you to think about all of the places that functional groups could come from so that if you get into a bind you can pull a different tool out of your toolbox. One more tool for that box is the Corey-Chaykovsky reaction. Don't worry about remembering that name; just remember that you can convert ketones and aldehydes to epoxides using this reaction. Below is a more detailed description of where the "CH_2" in the new epoxide comes from:

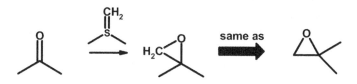

Here are several quick examples of the reaction:

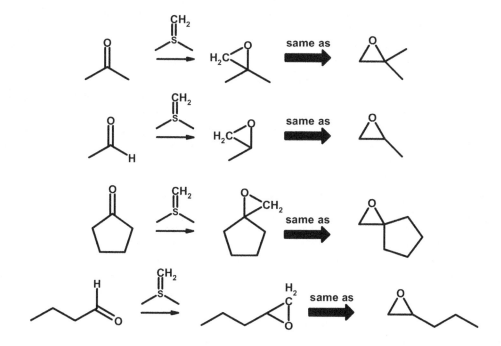

The reagent shown to perform these reactions is a sulfur ylide. The mechanism for the reaction is beyond the scope of this text, but suffice it to say that it is very effective at installing terminal epoxides in molecules. The CH_2 at the end of the sulfur ylide is effectively installed in between the C=O bond to give an epoxide. This means now that you would have two ways that you could create the epoxide product shown below:

Synthesize the following epoxide from any alkane or alcohol

Based on what we have learned to this point, there are two methods you could use to get full credit on this problem:

Method 1:

Method 2:

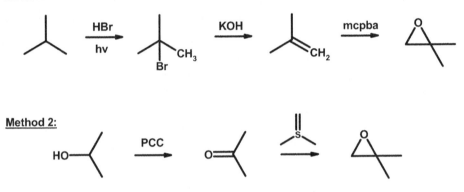

Method 1 is the more standard sophomore organic chemistry method, using halogenation, elimination, and epoxidation under standard mcpba conditions. Method 2 uses the Corey-Chaykovsky method, as shown above. Both of these synthetic methods are valid and should get you full credit. Becoming comfortable with more than one way to synthesize a molecule will help you to be better prepared for problems you have never seen before.

Take Home Message: Epoxides are usually created using mcpba. The Corey-Chayovsky method is an alternative method for creating epoxides.

#76- Two ways to open one epoxide

Epoxides are electrophilic because of resonance. This means that they can be attacked by nucleophiles and opened to give substituted alcohols as shown below.

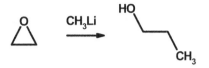

The example above is of a symmetric epoxide. What happens when the epoxide is not symmetric? Which side of the epoxide will our nucleophile attack? The answer is that it depends on what type of nucleophile is attacking. First, we will look at what happens when a basic or non-acidic nucleophile attacks an epoxide.

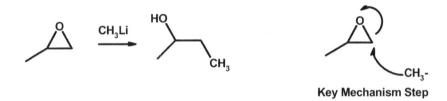

Key Mechanism Step

In this example, it is sterics that determines the outcome. Hence, the nucleophile will go to the LESS hindered side. Now, let's look at what happens under acidic conditions.

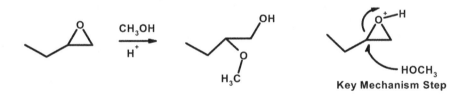

Key Mechanism Step

As we know, the first thing that happens in an acid-base reaction is the proton transfer. When we transfer the proton to the epoxide oxygen here, we have the chance to see some stable resonance structures. The most stable resonance structure here is going to be the one where the carbocation is on the most substituted carbon. This makes the substituted carbon the more electrophilic of the two and thus drives the nucleophile to the MORE substituted carbon under acidic conditions.

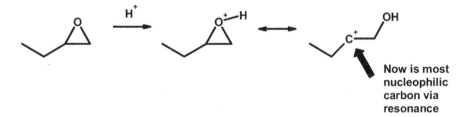

Now is most nucleophilic carbon via resonance

We can use this to our advantage and can determine which side certain nucleophiles attack by using the right reagents as shown below.

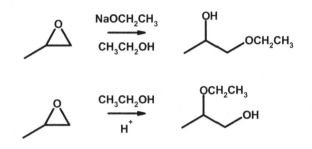

In this first reaction, we used basic sodium ethoxide to get the substitution at the less substituted carbon (steric control). In the second reaction, we used acidic ethanol to get substitution at the more substituted carbon (electronic control).

Take Home Message: Under acidic conditions, a nucleophile will attack an epoxide at the more substituted side as the reaction is under electronic control. Under non-acidic or basic conditions, a nucleophile will attack an epoxide at the less substituted side as the reaction is under steric control.

Chapter 5: Spectroscopy

#77- SODAR is not just a drink mixer anymore

Believe it or not, you can determine much more from a compound's molecular formula than just what atoms are in it. We can figure out how many rings and/or double bonds are contained in it too. This can be called the degree of unsaturation, or the Sum of Double Bonds and Rings (SODAR). In many of your spectroscopy problems, you will be given a molecule formula, and can calculate your total number of double bonds and rings in the molecule using the formula (2#C + 2 - #H - #X + #N)/2 where:

$\quad$ #C = the number of Carbons
$\quad$ #H = the number of Hydrogens
$\quad$ #X = the number of Halogens
$\quad$ #N = the number of Nitrogens

In using this formula, you ignore the oxygen or sulfur atoms. For example, the molecular formula C_6H_6NOCl would be (2*6 + 2 – 6 –1 +1)/2 = 4, meaning that there are 4 double bonds and/or rings. A double bond in this case can be an alkene or a carbonyl. Further, an alkyne counts as 2 double bonds. It is helpful to remember that benzene rings equal to 4 on the SODAR scale, (3 double bonds and one ring), so if you have a SODAR that is 4 or larger, think benzene ring somewhere in your molecule.

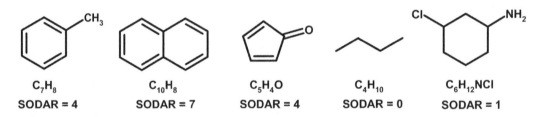

C_7H_8	$C_{10}H_8$	C_5H_4O	C_4H_{10}	$C_6H_{12}NCl$
SODAR = 4	SODAR = 7	SODAR = 4	SODAR = 0	SODAR = 1

Once you have calculated the SODAR for the compound, you can start on the other steps for examining the spectra you have been given to determine the structure of the unknown compound.

Step 2: Look for arene protons in the ^1H NMR. The number of protons between 6.5ppm-8ppm, known as the AR region, can give many clues to your molecule. We will discuss this aspect more in the next tip.

Step 3: Look for the 2 "A's", aldehydes and alcohols in your ^1H NMR. This is actually simpler than it sounds, and can give you some nice clues. Aldehydes are sharp singlet peaks that show up past 9ppm. Alcohols are broad singlets that can show up anywhere in the spectrum, but will "exchange" with D_2O, meaning that they will disappear if D_2O is added. Most organic chemistry professors will signify this by writing "exchange" over your spectrum.

Step 4: Add up the integrations in your spectrum and make sure it equals the number of protons that you have. For example, if you have 10 H's in your formula, but can only have an integration equal to 5 on your spectrum, you need to realize that each integration is equal to 2 protons.

<u>Step 5</u>: Start to make fragments and then add up the fragments. Using the integration and splitting of each peak, you can start to write down fragments of the molecule. For example, if you have a singlet with an integration of 3, you know that you have a methyl group (3 H's) next to something with no protons. If you have a doublet with an integration of 2, you have a CH_2 that is next to a CH. Once you have all of your fragments, start to piece them together and you will be figure out what your molecule is.

#78- Splitting should be automatic, at least with NMR

On the whole, undergraduate NMR problems are usually not too too difficult. As we discussed in the previous tip, you give yourself a head start by determining the total number of rings and double bonds in a molecule. In this section, we will show you some common splitting patterns for alkyl groups in an NMR. If you recognize these splitting patterns quickly, and can determine which alkyl fragment they came from, you will be well on your way to solving the problem quickly.

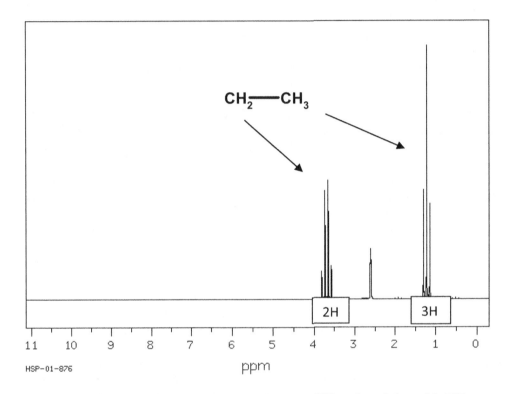

HSP-01-876

The ethyl group (CH₂CH₃): **a quadruplet with 2H and a triplet with 3H**

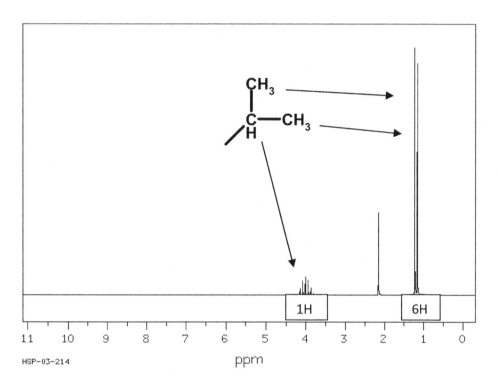

HSP-03-214

The isopropyl group [CH(CH₃)₂]: **a quadruplet with 1H and singlet with 6H**

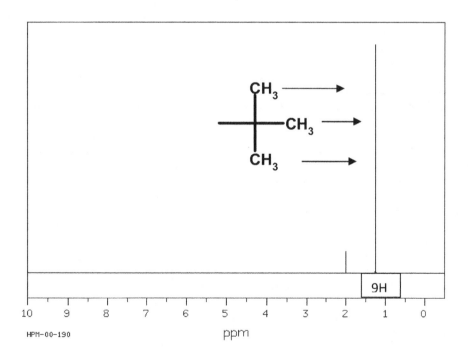

HPM-00-190

The t-butyl group: **a singlet with 9H**

Above are three sample proton NMR spectra designed to show a point. Each one of those three groups has a specific splitting pattern and should be able to be recognized very quickly. Each ethyl substituent will always be represented by a quadruplet and a triplet, having two and three protons, respectively. An isporopyl substitutent will always have a quadruplet/mulitple representing the CH, and a singlet representing the two methyl groups. One note here is that sometimes there will be two singlets in the spectra, as shown above. This will happen if the molecule cannot rotate quickly in solution, which causes the two methyls to show up as distinct. Therefore, with the isporopyl group, you could either have one singlet with 6H or two singlets with 3H each. The t-butyl is group is much more simple though, as it just has one singlet with 9H, representing all three methyl groups.

#79- O, M, P from the aromatic region of an NMR

When interpreting a proton NMR, there are few things that happen north of 6.5ppm. However, those peaks that are observed at shifts greater than 7ppm are significant and should be addressed. First, and most importantly, all aromatic protons will show up at 6.5ppm or greater. Thus, we can easily tell how substituted our aryl ring is by the number of protons in the aromatic region. The complex formula for determining the # of substituents on the ring is:

of aryl substituents = 6 − [# of aromatic protons]

Hopefully this formula did not blow your mind. Thus, if there are five aromatic protons in your spectra, you have a mono-substituted aryl ring. If there are three aromatic protons, you have a tri-substituted aryl ring.

More interestingly, if there are four aromatic protons, it means that you have a di-substituted aryl ring. But where are the substituents located in reference to each other? If you have a para substituted ring, it is a little easier to see than if it is the other two. As you can see below, para substituted rings give very clean aromatic regions, with two doublets between 6.5-8ppm. This is because there are only two UNIQUE aryl protons: the two closer to X and the two closer to Y.

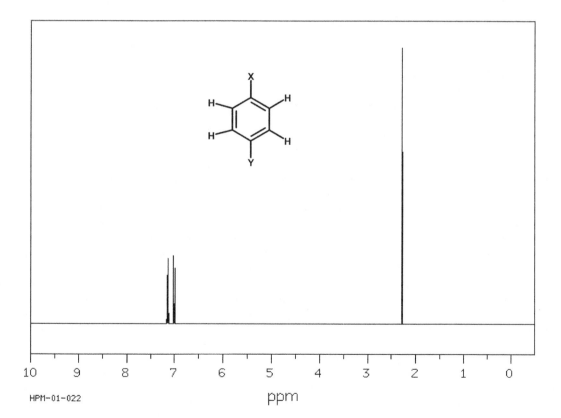

HPM-01-022

ppm

It is a little more difficult to tell when you have an ortho or meta substituted ring however. Theoretically, a meta substituted ring should have a singlet in the aromatic region, for the ortho proton which is not adjacent to any other protons. However, in reality, this singlet is usually obscured with the other aromatic protons and cannot be distinguished. Therefore, it is not prudent to rule out a meta-substituted ring just because you can't see the aromatic singlet.

Other than the aromatic protons, the only peaks that should show up at shift greater than 8ppm are aldehyde and alcohols (sometimes). Aldehydes will have peaks that are very sharp singlets, representing the proton attached to the carbonyl, located at shifts greater than 9ppm, depending on the molecule. The OH proton of an alcohol can appear in a variety of places, depending on the concentration of the solution. Many times, the OH proton is a broad singlet and can appear at shifts greater than 8ppm on occasion.

Take Home Message: **The aromatic region (6.5-8ppm) can tell us a lot about a molecule.**

#80- Everything you ever wanted to know about a ^{13}C NMR, but were afraid to ask

I guess ^{13}C NMR is kinda like the yin to ^{1}H NMR's yang. The two complement each other and give information which is different, but equally important. Truth be told, ^{13}C NMR is much easier to interpret than a proton. There are really two things you want to look at in each ^{13}C NMR that you run across:

1)   <u>How many peaks are there?</u> Each peak in a ^{13}C represents a unique carbon atom in your molecule. Therefore, if you have 8 peaks, you probably have an 8-carbon molecule. I say "probably" because symmetry needs to be taken into account. There is a very good chance that equivalent carbons, such as the methyl groups on a t-butyl group, will give one signal and therefore be indistinguishable. So, remember that the number of peaks represents the number of **_unique_** carbons, not necessarily to total number of carbons.

2)   <u>Where on the spectra are the peaks?</u> ^{13}C, more than ^{1}H NMR, has certain carbon atoms' signals fall into very specific ranges. Below is a generic chart of these. The most important peaks to quickly recognize are the alkyl carbons (generally 0-60ppm), aromatic carbons (130-160ppm) and the carbonyl carbons (180-220ppm). Unlike in ^{1}H NMR, you will see 6 aromatic carbon peaks **_every time_** you have an aryl ring because ^{13}C NMR doesn't care if there are protons bound to the carbons or not; it shows all 6 aryl carbons regardless.

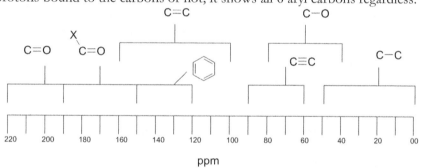

What will a carbon NMR not tell you? Unlike proton NMR, carbon NMR cannot tell you what the carbon is next to, as there is not splitting in a decoupled carbon NMR. Also, a decoupled carbon NMR cannot tell you how many protons are on the carbon, as there is not integration like in a proton NMR. Thus, care must be taken not to over-interpret the carbon NMR.

<u>Take Home Message</u>: Two things are important from carbon NMR: how many carbons there are and what type of carbon they might be.

#81- There are only four important IR peaks.

Amazingly enough, IR is not used much by professional organic chemists. This is because all IR can show is different functional groups. Thus, IR will have trouble telling the difference between any of the molecules shown below:

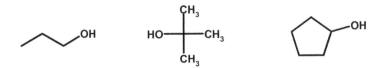

All of the spectra will show an OH peak and various C-H stretches, but each spectrum will look striking similar. Hence, we should recognize the limits of the instrument and not try to use it for more than it is intended. Further, there are really only four peaks in an IR that we look for to tell us something about our unknown molecule. Here, we show them in order of importance.

1) Carbonyl peak (1750-1650cm⁻¹): This will be a very sharp, prominent peak and shows that a carbonyl is present in somewhere in your unknown molecule. What is less obvious is which type of carbonyl it is. It is not as easy to distinguish between ketones, esters, aldehydes, etc.

2) OH peak (3500-3200cm⁻¹): This is a large, broad stretch which cannot be mistaken for any other functionality. One problem to be aware of is that the OH of water will also show up here, in the event that your unknown is not totally dry. Remember, that this can be from an alcohol OH or a carboxylic acid OH.

3) C-O peak (1300-1040cm⁻¹): Usually a large, sharp peak, this can be from an alcohol, carboxylic acid, ether, or an ester.

4) C≡N (2250-2230cm⁻¹) and C≡C (2100-2280cm⁻¹) peaks: Usually rather small peaks, but easy to spot as they are the only peaks in that area.

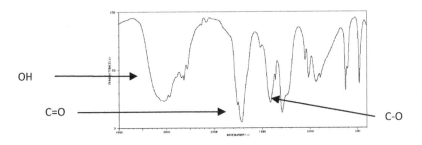

The biggest deal here is to recognize that the IR has its limitations and to not try to get more information out of it than it actually has. It would be a mistake to over-interpret the IR and draw misleading conclusions from it.

Take Home Message: Know the four peaks to look for in an IR and don't over interpret it.

#82- Check out the cleavage on that molecule

As we all know, mass spectrometry (commonly referred to as "mass spec") is a method for molecular identification which involves breaking up a molecule and then detecting the weight of its fragments. More specifically, a molecule is subjected to a high-energy electron burst, which can cause the molecule to become a radical cation, also known as a molecular ion (M^{+*}). This can then fragment in several different ways:

1) <u>Simple Cleavage</u>: The molecular ion loses a heteroatom, or other type of functional group. In this cleavage, both electrons from the cleaved bond are taken by the heteroatom X.

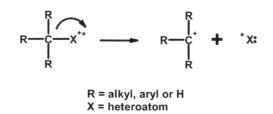

R = alkyl, aryl or H
X = heteroatom

2) <u>α-cleavage</u>: In this cleavage, the heteroatom remains on the molecular ion, and forms a double bond. (This mechanism is favored when X= O, S, or N). This cleavage proceeds via hemolytic bond cleavage.

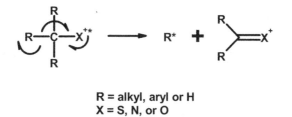

R = alkyl, aryl or H
X = S, N, or O

3) <u>Elimination</u>: This process is very similar to E2 elimination, and requires a free β-proton.

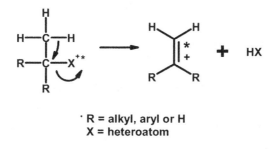

˙R = alkyl, aryl or H
X = heteroatom

Further, there is one type of rearrangement that is also a favorite of many organic chemistry professors. This is the McLafferty Rearrangement.

While this may look difficult, the rearrangement can be broken down into simpler parts. The first requirement for the rearrangement is a carbonyl somewhere in your molecular ion. Second, you must have a proton on the γ-carbon (the third carbon from the carbonyl). Once you have established that you have both of these, place your molecular ion in a bent configuration as shown above. This is the six-membered transition state that makes this a thermodynamically preferred transformation. Now that you have all of the parts in their proper place, draw your arrows and cleave the bonds so that you end up with a charged enol and a double bond. Congratulations, you have done your first McLafferty. This rearrangement is very common in mass spec and is seen on a number of substrates, as shown below.

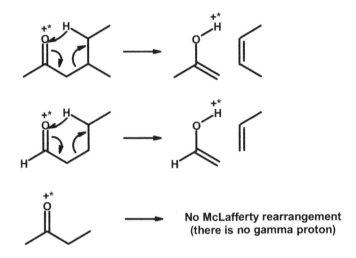

Take Home Message: There are three major mass spec cleavages. You need a carbonyl and a γ-proton to have a McLafferty rearrangement.

#83- The nitrogen hint (not a rule)

Just like with some of the other "rules" we have heard of, the nitrogen rule is not a real rule either. This is merely a guide, or a suggestion, to help us out with the number of nitrogen atoms that MAY be in a molecule.

The "rule" states that if you have a normal organic molecule, and the molecular weight is an even number then you have an even number of nitrogen atoms in that molecule. Conversely, if you have a normal organic molecule and the molecular weight is an odd number then you have an odd number of nitrogen atoms in that molecule. So, what is a normal organic molecule? It is a compound that contains only C, H, N, O, Si, N, S, and P. Some examples are shown below:

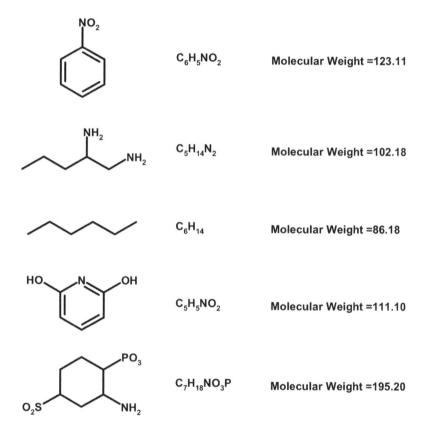

$C_6H_5NO_2$	Molecular Weight = 123.11
$C_5H_{14}N_2$	Molecular Weight = 102.18
C_6H_{14}	Molecular Weight = 86.18
$C_5H_5NO_2$	Molecular Weight = 111.10
$C_7H_{18}NO_3P$	Molecular Weight = 195.20

Where can you use this to your advantage? The rule is most helpful when dealing with a mass spec problem. Thus, if a parent ion in a mass spec problem is an odd number, then you most likely have a nitrogen atom somewhere in your unknown molecule. I say "one" nitrogen atom because it is unlikely that you will run across a molecule in your undergraduate courses that has three or more nitrogen atoms, therefore it is most likely that it will only be one if you have an odd mass.

One final point on the "nitrogen hint": It only applies to organic molecules, not inorganic ones. Therefore, your compound must have carbon and hydrogen in it somewhere. As an example, the nitrogen rule does not work for NO and NO_2.

Take Home Message: In a mass spec problem, the best application for the nitrogen rule is that if the parent ion's mass is odd, you most likely have one nitrogen atom in the molecule.

<u>Chapter 6</u>: Study Tips and Suggestions

#84- Are You a Learner Like Socrates or a Memorizer Like a Super Computer?

It has been studied, mused over, and contemplated for centuries. Which style of learning is better? Is someone who learned the "why" more educated or knowledgeable than someone who just memorizes the "what"?

It may be surprising to you that we are not going to say which is better. Both have their merits, and both can get you an "A" in undergraduate organic chemistry. What we are going to try to do is help you determine which one you are. Different situations, different circumstances, and different professors will lend themselves to one technique over the other.

From a theoretical standpoint, if you had an infinite amount of studying time, we would recommend that you do both learn and memorize as much material as possible. This way you can spit out an answer quickly when necessary and at the same time be able to reason your way through a new problem that you might never have seen before. However, with biochemistry courses, physics classes, and Dancing with the Stars, who has time to do both? Here, we hope to help you decide which you would like to do. Therefore, ask yourself these three questions:

1) Which do you feel more comfortable with? We have all been in school long enough by this point to know which we have a preference for. Some people just crave knowledge and cannot just accept that they just need to memorize something. Others just want to know that fact and move on to memorize the next fact. Deciding on your personal preference will go a long way to deciding whether you should memorize or learn the material for this class.
2) Are you going to continue in organic chemistry or is this the only time you will use this material? If you are not ever going to use organic chemistry again, then it is much easier to justify memorizing the material. If you are planning to go to medical school or take more chemistry courses then it will be a disservice to not learn the material, as later on you may need to have a good grasp on this material before venturing into tougher tasks.
3) Does your professor use the same exam questions over and over again? If so, then memorizing might not hurt you too much. If your professor thinks of new problems each exam, or makes you reason your way through a problem you have never seen before, then you probably want to be learning the materials.

You are going to be the only person who can determine which method is best for you. Now that you have determined that method, what is the best way to study based on your preference? Time and time again, we have seen that there is a simple answer to this: If you are a memorizer, flash cards are your best study aide. If you are a learner, try to learn as many reaction mechanisms you can.

#85- What a tangled web we weave.

Well webslingers, we are at the end and we have found that this is a wonderful study aide for students at almost any time in the semester. All it entails is creating a "link chart" between functional groups. Start with a simple alkane and think of all of the reactions you know on alkanes. Then take alkyl halides and figure out all of the reactions you know for those. Pretty soon you will see that you are finding multiple methods to get to the same functional group and multiple reactions you can do starting with a certain functional group. Your lines will start to cross a lot and you will see that all of the functional groups are interrelated. This is the essence of functional group interconversion (FGI), and is a very useful tool to master. Below, we have placed a sample. Be sure to place conditions for each reaction somewhere in the web while you are doing it. Your web should be different from ours based on the different reactions which are important to your class. The nice thing about this it is will help you study and simplify synthesis problems.

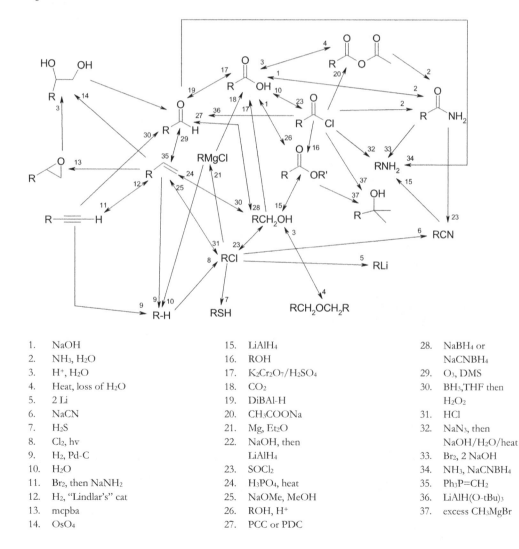

1. NaOH	15. LiAlH$_4$	28. NaBH$_4$ or
2. NH$_3$, H$_2$O	16. ROH	NaCNBH$_4$
3. H$^+$, H$_2$O	17. K$_2$Cr$_2$O$_7$/H$_2$SO$_4$	29. O$_3$, DMS
4. Heat, loss of H$_2$O	18. CO$_2$	30. BH$_3$,THF then
5. 2 Li	19. DiBAl-H	H$_2$O$_2$
6. NaCN	20. CH$_3$COONa	31. HCl
7. H$_2$S	21. Mg, Et$_2$O	32. NaN$_3$, then
8. Cl$_2$, hv	22. NaOH, then	NaOH/H$_2$O/heat
9. H$_2$, Pd-C	LiAlH$_4$	33. Br$_2$, 2 NaOH
10. H$_2$O	23. SOCl$_2$	34. NH$_3$, NaCNBH$_4$
11. Br$_2$, then NaNH$_2$	24. H$_3$PO$_4$, heat	35. Ph$_3$P=CH$_2$
12. H$_2$, "Lindlar's" cat	25. NaOMe, MeOH	36. LiAlH(O-tBu)$_3$
13. mcpba	26. ROH, H$^+$	37. excess CH$_3$MgBr
14. OsO$_4$	27. PCC or PDC	

#86- Be a Chatty Patty and Talk Out Your Reactions.

We all know about those people who just won't shut up. Whether it is in lecture, at a bar, or in a library, some folks just won't shut their pie hole. If you are one of these people, use it to your advantage. We have found that talking out organic chemistry reactions can make them easier to learn. When you do this, it is very helpful to use all of the lingo you have learned in class. For example:

Reaction:

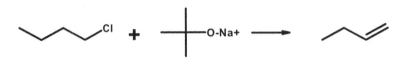

Spoken as: "This is a reaction between 1-chlorobutane, a primary alkyl halide and sodium tert-butoxide, a strong and hindered base. This reaction will proceed via an E2 mechanism to give 1-butene as the major product."

Reaction:

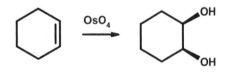

Shouted out loud as: "This is a strong oxidation of ethanol to acetic acid using the Jones reagent."

Reaction:

Preached to the people as: "This is a dihydroxylation reaction between cyclohexene and osmium tetroxide. This gives the cis diol product."

Take Home Message: Talking out your reactions, just like talking out your problems, will make you feel much better and help you to learn organic chemistry that much quicker.

Bonus: So, I have survive/passed/enjoyed this organic chemistry class...now what?

A presumptuous congratulation on passing this class. For some of you it was not easy. For others, it might have been enjoyable. And for a select few of you, it was so awesome that you would love to take the class again just for fun. Right now, I am talking to that last group of students.

If you enjoyed your organic chemistry class, you might want to consider a way to continue the party. That way is to get a graduate degree in chemistry. Here are some of the benefits to it:

1) Recruiting trips: Rent the 80's hit movie "Johnny Be Good". Your recruiting trips to prospective graduate schools will not be quite that crazy, but each school you are accepted to will fly you out for the weekend to wine and dine you. This includes meeting the faculty & current graduate students, seeing the campus, hearing about research that you might be interested in and seeing what life as a grad student would be like. It is a great way to spend part of your senior year and is the first step to picking the perfect graduate school for you. It is also an amazing opportunity to talk to the graduate students that are already there and find out how life at that school really is.

2) You get paid to go to school: Almost every university that offers a graduate degree in chemistry will pay you go to school there. No joke. In exchange for teaching undergraduate classes and/or doing research in order to obtain your degree, these schools will pay you a stipend. Generally, it is not much money, but it will be enough for most of you to live on. Depending on the school, this stipend can range from $15K to $35K/year and tuition is usually covered in that (or is very cheap). Considering that you are being paid to be a student, this isn't such a bad deal.

3) You get to put off starting real life: If you get a master's degree, it will take you 18 months to three years to complete. If you get a PhD, it will take you between 4-6 years. This is all time in which you are still a college student and can continue to party like it is 1999.

4) You will increase your earning potential for your entire career: With an advanced degree on your resume, you can demand higher salaries for your entire working career.

5) You don't necessarily even need to become a chemist with your degree: A sizable percentage of those who get advanced degrees in chemistry never actually become bench chemists, or even stay in the field of chemistry. I know people that have become engineers, pharmaceutical sales reps, medical examiners, and even FBI agents. The great part about it is that you have flexibility and aren't pigeon-held into a chemistry job.

Overall, more education never hurts anyone, especially when someone else is paying for you to do it. If you are even remotely interested in hearing more about this, I would strongly suggest learning more about a graduate degree in the sciences. For most schools, you can visit their websites and get more information. If you decide to start the process toward going to graduate school, you want to take the GRE exam sometime in your junior year and start applying in the fall of your senior year.

Made in the USA
Monee, IL
08 January 2022

88406741R10096